AF815693

Supramolecular Structure and Function 7

Supramolecular Structure and Function 7

Edited by

Greta Pifat-Mrzljak
Rudjer Bošković Institute
Zagreb, Croatia

Springer-Science+Business Media, LLC

Library of Congress Cataloging-in-Publication Data

Supramolecular structure and function 7/edited by Greta Pifat-Mrzljak
p. cm.
Proceedings of the 7th International Summer School on Biophysics: Supramolecular Structure and Function, held Sept. 14–26, 2000, in Rovinj, Croatia—T.p. verso.
Includes bibliographical references and index.
ISBN 978-0-306-46672-4 ISBN 978-1-4615-1363-6 (eBook)
DOI 10.1007/978-1-4615-1363-6

1. Biomolecules—Congresses. 2. Supramolecular chemistry—Congresses. I. Title: Supramolecular structure and function seven. II. Pifat-Mrzljak, Greta. III. International School on Biophysics "Supramolecular Structure and Function" (7th: 2000: Rovinj, Croatia)

QP501 .S836 2001
572.8—dc21

2001038564

Proceedings of the 7th International Summer School on Biophysics: Supramolecular Structure and Function, held September 14–26, 2000, in Rovinj, Croatia.

ISBN 978-0-306-46672-4

© 2001 Springer-Science+Business Media New York
Originally published by Kluwer Academic / Plenum Publishers, New York in 2001

10 9 8 7 6 5 4 3 2 1

A C.I.P. record for this book is available from the Library of Congress

All rights reserved

No part of this book may be reproduced, stored in a retrieval system, or transmitted in any form or by any means, electronic, mechanical, photocopying, microfilming, recording, or otherwise, without written permission from the Publisher

Preface

This volume represents a collection of lectures delivered by outstanding specialists in the fields of biophysics and of related scientific disciplines during the 7th International Summer School on Biophysics held in Rovinj, Croatia from 14 to 25 September 2000 under the title *"Super molecular Structure and Function"*.

This scientific-educational event was organized by the Ruder Boskovic Institute of Zagreb, Croatia with substantial material and intellectual support of a number of national and international institutions including the Croatian Biophysical Society (CBS), the International Union of Pure and Applied Biophysics (IUPAB), the International Centre for Genetic Engineering and Biotechnology (ICGEB) and the UNESCO Venice Office – Regional Office for Science and Technology for Europe (UVO-ROSTE).

The seventh edition of the series of International Summer Schools on Biophysics, which was started in 1981, attracted more than 120 young researchers and post-graduate students coming from 27 countries of Europe, Asia, Africa and Latin America.

Twenty-five outstanding experts in pure and applied biophysics presented the most advanced knowledge of this very interdisciplinary area of science during their lectures and round tables. It was commonly acknowledge that the Summer School achieved great success and fully reached its objectives.

The success of the Rovinj Summer School was also due to the constantly growing attention being paid by scientific communities to younger generations of scientists, thanks also to the major outcomes of the World Conference on Science "Science for the Twenty-first Century: A New Commitment" held by UNESCO and ICSU in Budapest, Hungary in June 1999.

It is believed that the knowledge on the most recent and advanced developments in biophysics presented at the School will be of great help for young scientists, who are beginning their professional careers in current and future research work in their home countries. The School could also serve as an important case of the most appropriate modality of the sharing and use of scientific knowledge.

The Organizers of the International Summer School on Biophysics also thought that the publication of this volume and its distribution within the scientific community would also serve towards the objectives of expanding, sharing and providing easy access to scientific knowledge as it was proclaimed during the World Conference on Science.

The publication of the volume is due to the substantial financial support provided by the UNESCO Venice Office – Regional Office for Science and Technology for Europe (UVO-ROSTE) as well as by the intellectual efforts of Prof. Greta Pifat-Mrzljak from the Ruđer Bošković Institute – initiator of the series of International Schools on Biophysics and also Chairperson of the 7th School held in Rovinj in 2000.

Organizing Committee

Contents

1. Aspects of the Evolution of Biophysics: Polyelectrolytes and the Weizmann
Henryk (Heini) Eisenberg 1

2. Analytical Centrifugation: Looking at Aggregation in Free Solution
P. Jonathan G. Butler 13

3. Mass Spectrometry of Peptides and Proteins - Principles and Features of Electrospray/Ionization-Mass Spectrometry (ESI-MS) and Matrix-Assisted Laser Desorption/Ionization-Mass Spectrometry (MALDI-MS)
Kerstin Strupat and Wolfgang Metelmann 27

4. Adsorption Kinetics of Low Density Lipoprotein onto a Hydrophobic-Hydrophilic Gradient Surface
Vladimir Hlady 45

5. Fluorescence Spectroscopic Studies on Structure and Function of Lipolytic Enzymes
Albin Hermetter, Birgit Mayer, Hubert Scholze, Elfriede Zenzmaier, and Marion Graupner 63

6. Infrared Spectroscopy of Lipoproteins
Xabier Coto and José Luis R. Arrondo 75

7. New Approaches in Spin Labeling and Spin Trapping. Part One: ESR Studies of Local Chemical Environment
Valery V. Khramtsov 89

8. New Approaches in Spin Labeling and Spin Trapping. Part Two: NMR Detects Free Radicals
Valery V. Khramtsov, Lawrence J. Berliner, and Thomas L. Clanton 107

9. Free Radical Intermediates of Drugs and Xenobiotics
Lawrence J. Berliner and Hirotada Fujii 119

10. *In vivo* Detection of Nitric Oxide: Combining EPR and NMR
Lawrence J. Berliner and Hirotada Fujii 131

11. On Biological Information
Manfred Eigen 143

12. Accuracy *versus* Efficacy in Biological Processes
Miroslav Radman 147

13. Towards a Memory-Based Interpretation of Proteome Data
János Murvai, Kristian Vlahoviček and Sándor Pongor 155

14. Modelling DNA Structure from Sequence
Kristian Vlahoviček and Sándor Pongor 167

15. Nonlinear Dynamics in the Binary DNA/RNA Coding Problem
Mladen Martinis 185

16. High Density DNA and Protein Arrays, Generation and Applications
Dolores J. Cahill 195

17. Mechanisms of Synaptic Plasticity in the Brain
Krešimir Krnjević 203

Subject Index 229

Aspects of the Evolution of Biophysics: Polyelectrolytes and the Weizmann

HENRYK (HEINI) EISENBERG
Structural Biology Department, Weizmann Institute of Science, Rehovot 76100, Israel

To provide a haven for Jewish refugee scientists fleeing Nazi wrath as well as a laboratory for Chaim Weizmann, the Manchester chemist and Zionist leader, the Sieff Institute, devoted to chemistry and microbiology, was established in 1933 in Rehovot, Palestine. The Institute was small and modest consisting of a main building, an excellent library established round the nucleus of books previously belonging and donated by the Jewish German scientist Fritz Haber, whose name it carried, some sheds for industrial research and a small cafeteria. Next to it was situated the Agricultural Research Station already established in 1921, to provide basic research and testing facilities for the dedicated tillers of the then barren soil.

The driving force leading the fledgling Institute in its timid initial steps was Ernst Bergmann, an organic chemist previously of Berlin, a devoted and faithful collaborator of Weizmann. During World War II the importance of the small research Institute gained considerably as a result of the services it was able to provide to the Allied Forces, cut off from regular channels of supply of some critical items. When Nazi Germany was vanquished and peace restored the future looked bright. The idea was conceived to present Weizmann, who was also the first President of the newly born State of Israel, with a birthday gift, a modern broadly oriented research Institute, bearing his name.

In June of 1946 a scientific symposium was held in Rehovot, to commemorate laying of the cornerstone for what was planned to become the "Institute of Physics and Physical Chemistry". A little volume "Trends in Modern Science" (1949) summarized the conference. In the preface Weizmann states that this was "the first time that a group of "Palestinian" (!)

Supramolecular Structure and Function 7, edited by Pifat-Mrzljak.
Kluwer Academic/Plenum Publishers, New York, 2001.

scientists met a group of distinguished visitors from abroad", hoping that these interchanges would be repeated as the "Palestinian" scientists had both much to offer and much to absorb. In his words, written in December 1948, Weizmann believes that "only science can lead the world back to sanity; the scientists and scientific institutions educating coming generations are responsible for the direction in which our civilization will develop". Also, from now on the Jewish student, the Jewish scholar, after having been a guest everywhere, a host nowhere, could finally be host to Jewish and non-Jewish students and scholars. Among the participants at this founding meeting were Louis Fieser, Ernst Bergmann, Saul Adler, Judah Quastel, Bernhard Zondek, David Rittenberg, Ladislaus Farkas, Herman Mark, Franz Ollendorf, Chaim Pekeris and Yizhak Elazari-Volcani, indicating a healthy mixture of chemistry, biology, medicine, physical chemistry, polymers, physics, mathematics and agricultural science.

Though plans had been made a few years earlier, the creation of the State of Israel and new currents in the march of Science brought about an almost complete change in emphasis in the development of the Institute. The scientific committee formed in 1944 to plan the Institute of Physics and Physical Chemistry consisted of Bergmann, Fieser, W.P. Hohenstein, Mark, Pekeris, Quastel, Rittenberg and Kurt G. Stern. Hohenstein, Pekeris and Stern were planning to leave the U.S.A. and settle in Rehovot to lead the Polymer, Applied Mathematics and Biophysics Departments. Stern and Hohenstein did not take up their positions at the Institute. Pekeris arrived in 1949 and set up the Department of Applied Mathematics, whereas Polymers and Biophysics were entrusted to my teacher Aharon Katchalsky, and his brother Ephraim Katchalski-Katzir respectively. Ephraim became well-known for his pioneering studies on the polymerization of amino acids and the use of the resulting polypeptides in biology (Katchalski-Katzir, 1995). His original ideas and work attracted many collaborators from Israel and abroad. Michael Sela, Ephraim's student, extended the use of synthetic polypeptides into chemical immunology, and founded a department at the Institute carrying this name (Sela, 1998).

Early in the 1950's the new building, now called the Ziskind building, became the central feature of a much broader budding Weizmann Institute and a temporary, extended or permanent abode for a large number of activities hardly envisaged by the founding fathers. I have already mentioned Applied Mathematics, Polymers and Biophysics, the latter moving into the Ullman building in 1963. Experimental Biology, led by Isaac Berenblum, the noted cancer scientist, stayed briefly in it in its initial stages until moving into the Wolfson building in 1952. Israel Dostrovsky explored the mineral resources of the southern desert of our country, the Negev, exploited the use of isotopes in chemical kinetics, and set up processes of water distillation to isolate oxygen isotope ^{18}O-rich water, a product still very much in demand today. Isotopes moved into the new building of Nuclear Sciences in 1958. Joe Jaffe headed the Infrared Optics laboratory which came into full swing with the operation of its giant infrared spectrometer in the basement of the Physics building in 1958. Gerhard Schmidt pioneered X-ray crystallography and solid state chemistry, and eventually moved in 1964 into the enlarged Sieff building, the Bergmann Institute. Benny Volcani, who had discovered the existence of life in the Dead Sea in the late 1930's in the form of a variety of microorganisms, continued for a while the tradition of the Sieff Institute in microbiology, and left for the States in the 1950's. Intensive activity was engendered by Pekeris who moved into geophysical research, search of oil and construction of digital computers which were a novelty in the world and unheard of in Israel. They were still running on electronic tubes before the days of the transistor and the microchip, and huge air-conditioning facilities and raised floors had to be constructed to provide the cooling capacity for the Weizac and the Golem computers.

I will, in the following, restrict my attention to the activities of the Polymer Department, with which I have been associated from its beginnings in 1949. I will emphasize a number of aspects, which made it so valuable to us. Its name is purely historical, but its activities transcended the limited significance its name might imply. Yet the concept of a polymer, or a macromolecule, is basic in the creation of life, be it on functional or on structural grounds, and it is no accident that the description of the DNA double helix structure by Watson and Crick in 1953 also emphasized its fundamental informational nature. It created the science of molecular biology, which has dominated modern science ever since.

The unique shape and content which the Polymer Department assumed owed much to the presence of Aharon but also to the inspired group of young students who gathered around him (Figure 1). Already at the Hebrew University in Jerusalem, and although trained in botany and zoology on Mount Scopus, in a University recently founded and still very limited in scope, he visualized a new class of charge-carrying polymers, polyelectrolytes, broadly significant in both biology and in the physical sciences. Initial studies of Aharon with Max Frankel were devoted to the interaction of amino acids and peptides with sugars in aqueous solution, leading to Aharon's doctoral thesis in 1939, at the young age of 25. Later Aharon supervised Nathan Sharon's doctoral thesis (1953) on aldose amino acid interactions. Nathan went on to lifelong intensive studies of carbohydrates and proteins and lectins (Sharon, 2000).

The name polyelectrolytes had been coined by Ray Fuoss in 1947, and in 1951 Ray became the first distinguished scientific visitor to visit the

Figure 1. Aharon Katchalsky, members and visitors to the Polymer Department at the front entrance of the Ziskind Building, Rehovot, 1965. Photo courtesy of Weizmann Institute Archives.

Polymer Department, attracted by our initial publications in this exciting new field. Structure in biology is heavily dependent on proteins, carbohydrates and nucleic acids, and diverse aspects of function relate to electrostatic charges spread on these macromolecules. This represented a basic and far-reaching thrust into the big unknown and provided tremendous stimulation towards an early understanding of biological structure and function, before X-ray crystallography, nuclear magnetic resonance, protein sequencing and many other superb tools of modern science became available. Werner Kuhn and Paul Flory had worked out the rules for the behavior of macromolecular chains in solution due to Brownian motion, and could relate these to hydrodynamic and thermodynamic manifestations, and sedimentation in the ultracentrifuge. Alex Silberberg came to us from Kuhn's laboratory and extended his studies on theoretical and experimental aspects of polyelectrolyte behavior. Alex and his students early discovered an interesting phenomenon distinguishing polymethacrylic acid, our major polyelectrolyte material in these days, from polyacrylic acid, which had been studied by Werner Kern, a student of Herman Staudinger. When solutions of polymethacrylic (not polyacrylic) acid were stirred rapidly, clear solutions gelled into solid-like material which would recover their liquid form after a while, demonstrating an important structural principle related to hydrophobic interactions deriving from the additional methyl group absent in polyacrylic acid. It was given the name negative thixotropy by the authors. My doctoral thesis (1952) guided by Aharon, on polyelectrolytes, molecular and macroscopic systems, emphasized the transition from polyelectrolyte solutions to mechano-chemistry, the transformation of chemical energy into mechanical work, resulting from crosslinking of molecular polyelectrolyte chains. This then provided the basis for my future work (Eisenberg, 1990a,1990b).

In my post-doctoral studies with Fuoss at Yale University in 1952 I would deepen my understanding in electrolyte behavior and statistical mechanics. Indeed, the Chemistry Department at Yale, with Kirkwood, Onsager, Fuoss, Sturtevant, Harned and Owen was the most sophisticated institution then dealing with these problems. Upon returning to the Polymer Department in 1953 I extended my work on simple electrolyte conductance performed during my stay at Yale, to the design of conductance cells useful in the study of ion condensation in polyelectrolyte solutions.

The first method which Aharon and Pnina (Spitnik) Elson had actually used in polyelectrolyte research was potentiometric titration. It is quite clear that as protons are removed from a polycarboxylic acid, an electrostatic field arises, which increases with increasing dissociation (pH). It becomes increasingly difficult to remove protons, the acid gets weaker and weaker, and the potentiometric titration of the polyacid is different from that of a simple acid: that is the theory in a nutshell. For further advances consult later work (Biophysics and other topics, 1976).

Great ideas in science and other human activities must be viewed in the context of the times in which they were created. In final analysis specificity and fine tuning are responsible for the precise functioning, preservation and replication of the biological machine, yet an understanding of the basic principles grafting function on structure cannot be easily dispensed with. Thus, early steps comprised physical analysis in solution, strongly affected by electrostatic charges, hydrogen bonds, and related factors creating difficulties in the determination of molecular weights, shapes of expanded "worm-like" coils, conductance, potentiometric behavior, surface interactions in solutions - studied by Israel Miller in the Polymer Department - and suspensions of natural and synthetic macromolecules. Viscosity was strongly shear-dependent in solutions of polyelectrolyte chains expanded by the repulsion of electrostatic charges, and we created special devices and approaches to overcome these problems. Classical rules devised to interpret the behavior of non-ionic macromolecules in the analytical ultracentrifuge, or the scattering of light or of X-rays were extended by Ed Casassa and myself, during my stay between 1958 and 1960 at the Mellon Institute in Pittsburgh then headed by Flory, by devising a theory for the thermodynamics of charged synthetic polyelectrolyte and biological multicomponent systems (Eisenberg, 1990a, 1990b).

Early excitement generated by the "kibbutznik" Shneior Lifson in developing theoretical approaches to understand the precise details of polyelectrolyte behavior was dampened by the realization of the enormous difficulties, persisting today, facing the solution of this problem (Lifson, 1997). No good theory was available, and is not available to the present day, for the determination of polyelectrolyte dimensions. Theories based on the expansion of coils all overestimated the force of repulsion and yielded fully stretched macromolecules, which was not confirmed from simple experiments (Eisenberg, 1976). Shneior and Aharon started to work on a model which, they felt, though it would not be useful to determine dimensions, would be useful for calculating potentiometric titration, ion-binding, conductance, osmotic pressure and so forth. This model states that for relatively short distances along the chain it is possible to assume a rod-like shape and calculate the distribution of counter-ions surrounding a cylinder with equally spaced oppositely charged co-ions. This model of polyelectrolyte behavior, later extended by Zeev Alexandrowicz to include simple salts, and known as the cell model, has maintained itself in one form or another to the present day. Already in 1963 Shneior, long the house theoretician of the Polymer Department, established the Department of Chemical Physics at the Institute, to deal with more spiritual aspects of the behavior of matter.

It was a well known and accepted concept in the early days of polyelectrolyte research that viscosity was an important method in the study of polyelectrolyte expansion with increasing charge and decreasing salt concentration. However viscosity was strongly shear dependent and the use of capillary viscometers was not leading to correct and reliable results. In a joint study with Jean Pouyet I was exposed to the Vallet Couette viscometer in Charles' Sadron Centre de Recherches sur les Macromolecules in Strasbourg and decided to design and build a Couette viscometer with electrostatic restoring force in the Polymer Department, covering a wide range of rates of shear, without using the classical delicate galvanometer suspension wire. I was fortunate to collaborate with Ephraim Heini Frei, then Head of our Electronics Department, on the original design of our precision rotation viscometer with electrostatic restoring torque, first reported at the founding session of the Israel Physical Society in 1954, and dedicated to Albert Einstein on his 75th birthday. In the construction of this instrument we also received valuable advice from our physicist friend Saul Meiboom, who later became one of the early pioneers in NMR research.

With the help of the instrument (Figure 2) we could determine viscosities of synthetic polyelectrolytes and of biological macromolecules such as DNA

Figure 2. Heini Eisenberg operating the Couette viscometer with electrostatic restoring force in the Polymer Department, Rehovot, 1955. Photo by Schleissner Photography, Rehovot.

and RNA, correctly extrapolated to zero rates of shear. Valuable information thus became available.

Polyelectrolyte chains connected in three-dimensional networks created, as already mentioned, mechano-chemical systems, capable of transforming chemical energy reversibly into mechanical work. Though far from leading to a correct interpretation of muscle contraction, which is still a hot topic of research to date, conceptual advances contributed to a better understanding of biological motion. Much interest was generated on the significance of polyelectrolyte behavior when the Watson-Crick DNA double helix was discovered, including subtle distinctions in the behavior of double and single-stranded polynucleotide chains. Next attention veered to irreversible thermodynamics, essential in the maintenance of the processes of life, and in particular to the behavior of biological membranes, in the work of Ora Kedem and Aharon (Figure 3).

Activities described above cut strongly across the disciplines of chemistry, physics and biology. Excitement was high and we were in the front-line of explosive research in a number of connected disciplines.

Continuous visual and audial exchange was generated around the blackboard in the Polymer Library. A continuous flow of visitors on short- and long-term visits led to frequent lectures in the overcrowded library filled to the brim, including occupation of the top of the bookshelves. Long-lasting associations were created across the continents, the big cupboards in the office filled with doctoral theses and enormous numbers of reprints. It was

Figure 3. Aharon Katchalsky lecturing on irreversible thermodynamics in the Polymer Department. Photo courtesy of Weizmann Institute Archives.

science at its best, with no artificial boundaries between the various disciplines. In the quest for knowledge and understanding of the laws of nature separation into divided compartments was meaningless (Eisenberg 1972).

In April 3-6,1956, we organized the first IUPAC International Congress in Israel, in Rehovot and in Jerusalem, on polymers and polyelectrolytes which did much to focus world-wide attention on our activities and interests. Outstanding scientists from the world over came to attend this meeting in a country which had come into existence only a few years before. To name just a few J.J. Hermans and J.Th.G. Overbeek came from Holland, Charles Sadron, Michel Magat and Henri Benoit from France, Paul Doty, Herman Mark, Ray Fuoss, Terrell Hill and Herbert Morawetz from the U.S.A.(Jim Watson of recent DNA fame was also there participating in the discussions), Arthur Peacocke from the UK, V.A. Kargin and V.N. Tsvetkov from the then Soviet Union, Michel Mandel and G. Smets from Belgium, Fumio Oosawa and A. Wada from Japan, Anton Peterlin from then Yugoslavia. Sadron brought his whole Laboratory from Marseilles to Haifa in an Israeli passenger boat. Little did they realize that this was the week of Pesah (Passover) and they were served only Matzot, no bread, on this trajectory. However this did not discourage their enthusiasm. The meeting was extremely successful and eighty papers were published in the J.Polymer Sci. in 1957 (IUPAC, 1957).

Involvement with polymers, with polyelectrolytes, with fibers and with membranes led to a number of applied activities and interactions with industry and agriculture. With funds raised from the US Government we set up a versatile facility for plastics testing, extrusion, molding and literature search. Saul Gassner manufactured novel plastic mini-devices, not yet available at that time, advised the fast growing plastics industry and developed invisible freely breezing plastic coatings to protect oranges and other fruit. Eventually in 1970 the Plastics Laboratory became the independent Plastics Department, under the guidance of David Vofsi.

Aharon moved more deeply into the problems of the origin of life - trying to mimic prebiotic synthesis by polymerizing with Mela Paecht amino acids and nucleotides by heterogenous catalysis on mineral surfaces, such as swollen montmorillonite clays, - and into philosophical aspects of science. He was killed on May 30, 1972, in the main hall of Lydda Airport, together with many other innocent passengers, by a Japanese, hired by an Arab terrorist group - returning home from a meeting with Manfred Eigen in Göttingen. The senseless death of Aharon was a heavy blow to Israeli science. However a broad range of activities continued unabated even though the Department was weakened by the departure of the membrane group to establish an independent Membrane Department, led by Ora Kedem, in 1974. Symposia in memory of Aharon continue to be organized at regular intervals (Eisenberg, 1977).

Additional ways of expression, communication and knowledge interchange, meetings in which members of the Polymer Department participated or which they organized were Edmund de Rothschild Schools in Molecular Biophysics, sponsored by Bernard Pullman, the Institut de Biologie Physico-Chimique in Paris, on biomembranes and intracellular structures, DNA and chromatin, and modern aspects of halophilism. Aharon Klug was a regular teacher and participant, Morton Bradbury, Roger Kornberg, Harold Weintraub, Gary Felsenfeld, Richard Dickerson, Ken Van Holde, Pete Von Hippel, Pat Dennis, Rainer Jaenicke, Janos Lanyi, Dieter Osterhelt, Hans Günther Wittmann, Wolfram Zillig, Joel Sussman, Ada Yonath and many others. Aharon was a founder of the International Union of Pure and Applied Biophysics (IUPAB), and we were able, with Israel

Pecht, to organize the very successful 9th International Biophysics Congress in 1987 24-29 August in Jerusalem (Eisenberg, 1988; Eisenberg and Navon, 1988).

Interests in the Polymer Department continued to range over a wide spectrum, in step with the advancing frontiers of biophysical sciences. A shift of emphasis gradually developed, progressing from the fundamental study of synthetic charge-carrying polyelectrolytes to the investigation of more complex biological macromolecules, enzymes from halophilic bacteria surviving at extremely high saturated salt concentrations in the Dead Sea, nucleic acids and chromatin (Eisenberg, 1990b). A highly sophisticated laboratory for small angle X-ray scattering, of major use in biology and chemistry, was established and conducted by Ellen Wachtel and Reuven de Roos. Our Department became a unique place in the scientific world in which joint ultracentrifuge, elastic and inelastic light scattering, X-ray scattering and neutron scattering experiments, by collaboration with Joe Zaccai in Grenoble, could be undertaken on important biological macromolecular systems.

Glycoproteins represent a major secretion of the mucous membrane, the study of which was actively pursued by Alex on a physiological, biochemical and physico-chemical basis. Another related field of their study was concerned with the biorheology of epithelial mucus, blood, fibrin clotting and platelet aggregation. The study of macromolecular structures and complexes, from the theoretical as well as the experimental points of view, and attempts towards understanding relations between structure and function remained major aims in the activities of the Department.

Polymer science, long a field pioneered by physical chemists such as Kuhn, Flory, Bruno Zimm, Walter Stockmayer, "Geheimrat" Mark, an old friend and supporter of the Department, and many others, underwent a rediscovery process in the hands of pure physicists, as exemplified by Sam Edwards and by Pierre-Gilles de Gennes. As a matter of fact de Gennes, invited by Shlomo Alexander to Rehovot in 1966 for a solid state physics meeting, was all excited having quite recently discovered the existence of polymer science and spent many exhilarating afternoons and evenings with Alex, myself and other members of our Department, to strengthen his grip on this fourth state of matter. The lecture he delivered at the physics meeting dealt with polymers, and not with Solid State Physics. Later, in his Nobel lecture, de Gennes stressed the impact of polymer science on biological phenomena. In 1977 Jacob Klein, product of the new Polymer Physics trend, joined the Polymer Department, and extended his novel approach in the evaluation of forces on the Ångstrom scale by synthetic polymers absorbed on mica surfaces in solution. He could also find new uses for the antiquated van der Graaf generator in Nuclear Physics, for the study of solid polymers in thin layers. Jacob became the last head of the Polymer Department. Sam Safran, expert in the theory of colloids and surfaces, came from the USA to join the Department. Jacob Anglister introduced multi-dimensional NMR protein structural studies into Israel. Ed Trifonov who came in 1977, coined the term, now universally accepted, of DNA 'bendability', and is expert on manifold aspects of the genetic code, a field he calls Gnomics. Ed is continuously surrounded by recent immigrants from the former Soviet Union and tries hard to create conditions suitable for their successful absorption into the Israeli life and culture.

In October 1991 the Department was disbanded in a broad organizational move at the Institute and its members were transferred into the two newly founded Departments of Structural Biology and Materials and Interfaces. However this is another story which I may narrate at another future occasion.

REFERENCES

Biophysics and other Topics, 1976, *Selected Papers by Aharon-Katchalsky,* Academic Press, New York.

Eisenberg, H., 1972, Macromolecules in Research, Biology and Practice. *Bull. Atomic Scientists* **28**: 56-63.

Eisenberg, H., 1976, Polyelectrolytes, Thirty Years Later, *Biophys.Chem.* 7: 3-13.

Eisenberg, H., 1977, Symposium Editor, Macromolecular Chemistry - 11, *The Third Aharon Katzir-Katchalsky Conference*, Jerusalem, Israel, 13-18 July 1975, *Pure and Appl. Chem.*, **46**: No. 2-4.

Eisenberg, H., 1988, *Biophysics- Theme and Variation:* 9th International Biophysics Congress Jerusalem, Israel 24-29 August, 1987. *BioEssays* **9**: No. 1, 34-35.

Eisenberg, H., 1990a, *Never a Dull Moment*, G. Semenza and R. Jaenicke (Eds.). Selected Topics in the History of Biochemistry: Personal Recollections III, Elsevier, Amsterdam, *Comprehensive Biochemistry,* **37**: 265-348.

Eisenberg, H., 1990b, Thermodynamics and the Structure of Biological Macromolecules, Rozhinkes mit Mandlen, *Eur. J. Biochem.* **187**: 7-22.

Eisenberg, H. and Navon, G., 1988, Editors. *Invited Symposium Papers*, 9th International Biophysics Congress IUPAB, Jerusalem, August 24-29 1987. *Biophys. Chem.*, **29**:

IUPAC, International Union of Pure and Applied Chemistry. 1957, *Symposium on Macromolecules*, Rehovot, April 3-6, 1956. *J. Polymer Sci.*, **23**:

Katchalski-Katzir, E., 1995, My Life in and Beyond the Laboratory, *Annu. Rev. Biophys. Biomol. Struct.* **24**: 1-29.

Lifson, S., 1997, Wandering in the Field of Science, G. Semenza and R. Jaenicke (Eds.). Selected Topics in the History of Biochemistry: Personal Recollections V, Elsevier, Amsterdam, *Comprehensive Biochemistry,* **40**: 1-55.

Sela, M., 1998, Probing into the Realm of Proteins and Immunity, *Protein Science* 7: 1653-1659.

Sharon, N., 2000, *Half a Century between Carbohydrates and Proteins*, G. Semenza and R. Jaenicke (Eds.), Selected Topics in the History of Biochemistry: Personal Recollections VI, *Comprehensive Biochemistry*, **41**: 391-448 .
Trends in Modern Science, 1949, Weizmann Institute of Science, Rehovot.

Analytical Centrifugation: Looking at Aggregation in Free Solution

P. JONATHAN G. BUTLER
MRC Laboratory of Molecular Biology, Hills Road, Cambridge, CB2 2QH, UK

1. INTRODUCTION

An increasingly common question in structural biology now is the state of aggregation of a protein or complex in solution. In other words, is it monomeric, or has it aggregated to form some oligomer which is essential for activity? One approach to answering this question has been the use of gel permeation chromatography, where a porous gel matrix is packed into a column and the size of the complex measured in terms of its elution volume. The column is calibrated for elution volumes by using a set of standard proteins, of known molecular mass, and the mass of the unknown is estimated from a semi-logarithmic plot of molecular mass against elution volume. There is a problem with this approach, namely that the elution volume is actually a function of Stokes radius, rather than molecular mass, and so the calibration is based on the, often unacknowledged, assumption that the molecular shapes of the unknown and standards will be similar.

An alternative approach is non-denaturing gel electrophoresis, again employing marker proteins for calibration of the distance migrated for known molecular masses. A further problem is the requirement that the electric charge be the same, in order to give a meaningful mobility. At one extreme, if the net charge were zero the unknown would not move in the gel, even though it might be small enough to migrate readily were there any charge to interact with the electric field.

An alternative physical technique is analytical ultracentrifugation, where the material is centrifuged in a special rotor and cells, which allow analysis of the sample distribution while the centrifuge is running. This instrument

Supramolecular Structure and Function 7, edited by Pifat-Mrzljak.
Kluwer Academic/Plenum Publishers, New York, 2001.

was originally developed by The Svedberg and his collaborators (*e.g.* Svedberg & Pedersen, 1940), although their air-turbine instrument was not convenient in use and the technique really only caught on when an electrically driven instrument, the Spinco Model E, was developed. Since then there have been a number of different instruments, but they have all been electrically driven. The distribution of the sample radially along the cell is analysed optically, by having a vertical axis of rotation and clear windows above and below the cell, so that a vertical light path is available through the cell. The sample wells in the cells are typically sector shaped, with vertical sides and the radius normal to the axis of rotation. This allows material to sediment down the cell without impinging upon the sides of the cell, so that it only starts to pellet onto the very bottom and does not interfere with the analysis of the remaining sample. Various optical techniques have been employed to image the cell, particularly Schlieren optics, interference optics and absorption optics. Nowadays the latter is the most commonly used, with a monochromatic light source and a scanning slit above a photomultiplier tube. By comparing the light transmitted through sample and reference sectors, with the latter containing only the buffer, plots of the optical density against radius are readily obtained.

There are two main types of experiment, known as sedimentation velocity and sedimentation equilibrium (see *e.g.* Schachman, 1959 for wider discussion). The first was the technique developed by Svedberg, where the rate of sedimentation under a known centrifugal field is measured, to give the sedimentation coefficient. The centrifugal field depends upon $r\omega^2$, where r is the radius and ω is the angular velocity. The rate also depends upon the molecular mass (M_r), the buoyancy of the aggregate and its diffusion coefficient (D), which is related to the molecular size and shape, as well as the viscosity of the solvent. The buoyancy comes in a term $(1-\bar{v}\rho)$, where $\bar{v}$ is the partial specific volume (the volume of solvent displaced by one molecule of solute) and ρ is the solution density; $(1-\bar{v}\rho)$ is therefore effectively the "residual specific mass", after allowing for the buoyancy of the molecule in the solution. Both $\bar{v}$ and ρ can be either calculated or measured (see below) and are, in any case, fairly constant for a typical protein or nucleic acid, but the diffusion coefficient again causes a problem and is difficult to measure. Sedimentation coefficients are therefore useful parameters, but mainly related to molecular mass by comparison with other well characterised samples, *i.e.* we are again back to comparison with standards as for gel methods.

In sedimentation equilibrium experiments, the sample is sedimented sufficiently slowly that it reaches an equilibrium state where the distribution against radius is constant, with the rate of sedimentation balanced by the rate of diffusion down the concentration gradient established to the bottom of the cell. In this situation, the diffusion coefficient cancels out and the distribution is dependent only upon the sedimentation conditions, the molecular mass and the buoyancy of the sample.

2. ANALYSIS OF SEDIMENTATION EQUILIBRIUM

Under the conditions just described, the distribution can be described by the equation:

$$\overline{M}_{w,app} = \frac{d\ln(c)}{dr^2}\frac{2RT}{(1-\bar{v}\rho)\omega^2} \tag{1}$$

where $\overline{M}_{w,app}$ is the apparent, weight average molecular mass, c is the concentration of macromolecule at radius r, R is the ideal gas constant and T is the absolute temperature (Schachman, 1959). Since all of the terms on the right hand side of the equation can be either measured or calculated, this allows an absolute determination of the molecular mass and hence the aggregation state.

Although this form of the equation has proved very useful in many cases, it has the problem that the buoyancy is considered as the partial specific volume, which is technically the molecular volume *in vacuo* and the density is that of the solution, which varies with solute concentration, *i.e.* with radius at equilibrium. This is often not a problem, as neither differs significantly and one can use simple, calculated values (Laue *et al.*, 1992) (or even assume $\bar{v} = 0.725\,\text{ml}/\text{g}$ and $\rho = 1.01\,\text{g}/\text{ml}$) in many cases. However, it does present a problem with some interesting cases such as a membrane protein solubilised with detergent, where the protein will have bound a number of detergent molecules to its hydrophobic surface.

The solution to this problem was given by Casassa and Eisenberg (1964), who introduced the concept of an apparent partial specific volume (ϕ') and showed that the basic equation becomes:

$$\overline{M}_{w,app} = \frac{d\ln(c)}{dr^2}\frac{2RT}{(1-\phi'\rho_0)\omega^2} \tag{2}$$

where ρ_0 is now the solvent density. Moreover, Casassa and Eisenberg (1964) also showed that:

$$\left(\frac{\partial \rho}{\partial c_2}\right)_\mu = (1 - \phi' \rho_0) \qquad (3)$$

where ρ is the solution density at weight concentration c_2 of the macromolecular component and constant chemical potential of all other components (*i.e.* under conditions of dialysis equilibrium against the pure solvent), ϕ' is the apparent partial specific volume and ρ_0 is the solvent density.

The density increment $\left(\left(\frac{d\rho}{dc_2}\right)_\mu\right)$ can be obtained by measuring the density of the solvent and also solutions which have been dialysed exhaustively against solvent, and subsequently determining the macromolecular concentration in the solution. Such direct measurement, for the calculation of the apparent partial specific volume with equation (3), has the advantage that the molecular mass calculated from equation (2) will correspond to the species whose weight concentration was measured (Casassa & Eisenberg, 1964). In other words, if the protein alone is measured in a solution of detergent solubilised membrane protein, the molecular mass will correspond to the protein alone, with no effect of the bound detergent.

The molecular mass can therefore be calculated as a function of protein concentration and, taking this together with the monomer mass, the likely state of aggregation be deduced. In favourable cases, it is even possible to obtain an estimate for the dissociation constant (K_d) (*e.g.* Poget *et al.*, 1999). However the picture is not always clear, particularly if the macromolecule shows non-ideality. This is typically described by the equation:

$$\frac{1}{\overline{M}_{w,app}} = \frac{1}{M_0} + 2Bc \qquad (4)$$

where M_0 is the "true" molecular mass (*i.e.* at concentration, c, equals zero, so with no effect of non-ideality) and B is the second virial coefficient. In favourable cases, it is possible to resolve the aggregate state of a non-ideal complex (Perrett *et al.*, 1999), but such analysis has the fundamental limitation that one is fitting a derived quantity, rather than the raw data, and the error will not be normally distributed.

The development of modern computers has enabled this problem to be overcome by direct non-linear least squares fitting of the original data by the equation which was previously used to derive equation (1), allowing the molecular mass to vary as a parameter in the fit (Johnson & Straume, 1994). This gives the equation:

$$A = \varepsilon c_0 \exp\left(\sigma\left(r^2 - r_0^2\right)\right) + \Delta \tag{5}$$

where A is the absorbance, ε is the extinction coefficient of the macromolecule, c_0 the concentration at reference radius r_0, Δ is the error in the baseline and $\sigma = \dfrac{M_{app}(1-\phi'\rho_0)\omega^2}{2RT}$, with M_{app} as the apparent molecular mass.

Equation (5) can be adapted to allow for appropriate variations in M_{app}, for example for a monomer/dimer equilibrium it becomes:

$$A_t = \varepsilon_1\left(c_0 \exp\left(\sigma\left(r^2 - r_0^2\right)\right) + \frac{2\left(c_0 \exp\left(\sigma\left(r^2 - r_0^2\right)\right)\right)^2}{K_d}\right) + \Delta \tag{6}$$

where M_{app} is now equivalent to the monomer mass, A_t is the total absorbance, ε_1 is the molar extinction coefficient for monomer and K_d is the dissociation constant. Similarly, one can derive the equation for non-ideal solutes, *e.g.* for a non-ideal monomer one gets the transcendental equation:

$$A = \varepsilon c = \varepsilon c_0 \exp\left(\sigma\left(r^2 - r_0^2\right) - 2B\left(c - c_0\right)\right) + \Delta \tag{7}$$

where c is the concentration at radius r. The equations to fit other, more complicated models can also be derived, but one must be aware that fitting more parameters, while likely to give a better fit to the data, is fraught with danger unless they are based upon a model known to be likely.

One problem with all such fitting is in deciding between the fits given by the different models. In this case, however, it is possible to use the residuals between the calculated and measured absorbance values at each radius value, plotted against radius, as a good assessment of goodness of fit. For an appropriate model, these residuals should not only be small but, more importantly, should be randomly distributed around zero. Any systematic differences are good evidence that the model is not appropriate (Johnson & Straume, 1994).

3. SEDIMENTATION EQUILIBRIUM MEASUREMENTS FOR PROTEINS WITH UNUSUAL APPARENT PARTIAL SPECIFIC VOLUMES

An unusual apparent partial specific volume for a protein usually arises from an abnormal solvation; either the protein is binding some diffusible component from the solution, to the exclusion of the solvent, or else the solvent (usually water) is being bound to the exclusion of salts, *etc.* in the solution. Either of these can result in a volume associated with each protein aggregate which has a markedly different density from free solution, as well as potentially from the protein alone. In such cases the most reliable method for determining the aggregation state is to measure the apparent partial specific volume directly.

I would like to discuss a number of examples where, for different reasons, we have found this to be valuable in enabling us to resolve the stoichiometry of the aggregation.

3.1 Glycoproteins: the pentraxins

One common type of protein, which are not composed simply of amino acids, are the glycoproteins. These contain varying amounts of carbohydrate bound at specific sites on the peptide backbone, with the added complication that there is usually variation in the detailed structure of the carbohydrate tails not only between sites but even at a given site. This renders the appropriate partial specific volume difficult to decide, unless one measures it for the given sample.

One family of glycoproteins upon which we have worked are the pentraxins. They are a highly conserved family of plasma proteins, which includes the constitutive protein, serum amyloid P component (SAP), and the acute phase protein, C-reactive protein (CRP) (Osmand *et al.*, 1977; Pepys & Baltz, 1983; Pepys *et al.*, 1978). A characteristic of these proteins, leading to the family name of the "pentraxins" is their characteristic 5-fold symmetry, so that they are either pentameric or decameric, with 2 pentamers stacked on each other. The only positively identified pentraxin from an invertebrate is the CRP from the horseshoe crab, *Limulus polyphemus* (Robey & Liu, 1981), and this is of increased interest due to *L. polyphemus* being regarded as a "living fossil", which is believed not to have evolved significantly in recent time.

One challenge of *Limulus* CRP was that it appeared to show 6-fold symmetry in the electron microscope, although the images were not very clear. This therefore raised the question whether this ancient pentraxin

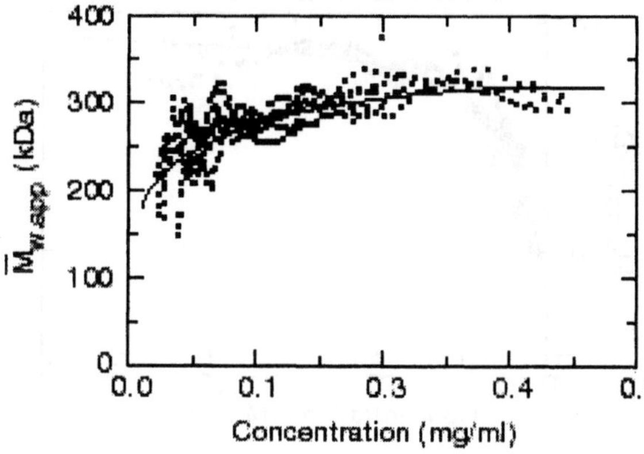

Figure 1. Plot of $\overline{M}_{w,app}$ against concentration of *Limulus* CRP, with data for multiple cells superimposed. The curve drawn is arbitrary, taken to fit within the data, as the polymorphic masses of the monomers prevent any precise curve-fitting to a model, although aggregation is clearly occurring up to ~12-mer.

might have a different symmetry from its more modern homologues. We therefore decided to investigate the size of the *Limulus* CRP in free solution by sedimentation equilibrium.

CRPs are glycoproteins, so we started by measuring the density increment, to calculate the apparent partial specific volume (Tennent *et al.*, 1993). A plot of density of the solution against CRP concentration yielded values of $\left(\partial\rho/\partial c_2\right)_\mu$ = 0.273±0.003 and $\rho_0 = 1.00726 \pm 0.00002$ g/ml, giving ϕ' = 0.722±0.003 ml/g. Because of the multigenic nature of *Limulus* CRP, together with its variable glycosylation, we had measured c_2 by refractometry and it was therefore the concentration of the glycoprotein; hence this fairly typical value for ϕ' applies to the entire glycoprotein and not just the peptide component. A plot of $\overline{M}_{w,app}$ against concentration (Fig. 1), for multiple runs with different sample concentrations loaded, shows aggregation from low protein concentrations to a molecular mass, substantially independent of concentration, of 300 kDa. Since the monomeric mass, based on electrospray mass spectroscopy, was 25–26 kDa, this shows that *Limulus* CRP has 12 subunits, *i.e.* that it is composed of dimerising hexamers, and confirms that the symmetry does differ from the other pentraxins which have been characterised.

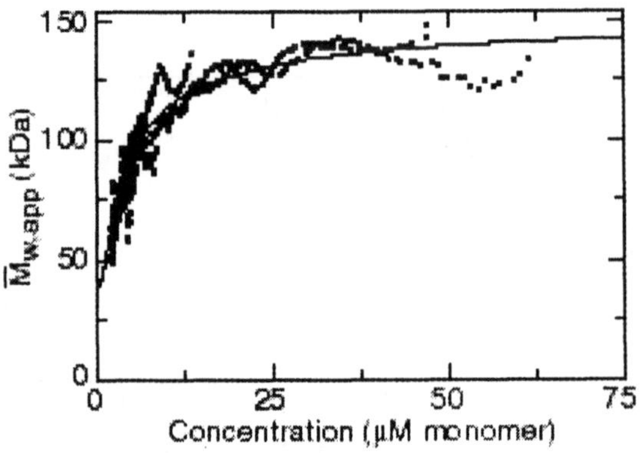

Figure 2. Plot of $\overline{M}_{w,app}$ against concentration of MIP from bovine eye lens fibre membranes. The curve is for a model of monomer/tetramer aggregation and was fitted to the data.

3.2 A membrane protein in detergent: the major intrinsic protein

Another common application is the investigation of membrane proteins which require detergent for solubilisation. A study which we made (König *et al.*, 1997) was of the major intrinsic protein from bovine lens fibre membranes (MIP). MIP forms ~50% of the protein in the mature fibre cell (Vermorken *et al.*, 1977), giving the eye lens its clarity which is essential for sight. We were working with MIP which had been solubilised with decyl-β-D-maltopyranoside. Measurement of the density increment, obtaining the protein concentration by amino acid analysis so that only the peptide part was measured with no effect of bound detergent, gave $\rho_0 = 1.0319$ g/ml and $\phi' = 0.734$ g/ml. Plots of $\overline{M}_{w,app}$ against concentration (Fig. 2) showed an association that was fitted by a monomer/tetramer equilibrium of the 28 kDa monomer. (Fitting with a monomer/dimer/tetramer equilibrium showed negligible dimer concentration.)

Electron microscopy showed a very thin, disk-like particle which had a 4-fold rotational symmetry when viewed normal to the disk, in keeping with it being a tetramer. The dimensions of the particle were also compatible with this conclusion. Incidentally, gel permeation chromatography had given an apparent molecular mass of 211 kDa which would correspond to

~98 kDa of detergent bound to the 113 kDa of protein, *i.e.* to ~200 detergent molecules bound per tetramer.

4. COMBINED SEDIMENTATION EQUILIBRIUM AND VELOCITY STUDY OF INHIBITOR IF_1 AND ITS BINDING TO THE MITOCHONDRIAL ATPase SUBUNIT F_1

ATP synthesis in the mitochondrion is driven by a proton-motive force which is built up by the transport of protons out of the mitochondrial matrix during oxidative electron transport. The mitochondrial F_0/F_1 ATPase then allows protons to re-enter the matrix, coupling this to the joining of ADP and inorganic phosphate by the F_1 subunits, i.e. synthesising ATP. Under anoxic conditions, electron transport stops and the proton-motive gradient across the membrane is lost, with the matrix pH falling to below pH 6.5. This allows F_1 to work backwards, as an ATPase, and hydrolyse cellular ATP. This adverse reaction is prevented by an inhibitor of F_1, known as IF_1, which binds to F_1 at lower pHs and inhibits the ATPase activity. We were interested to characterise IF_1 and also to see what we could find about its binding to F_1.

4.1 Sedimentation equilibrium measurements on IF_1

Because of the importance of pH in the activity of IF_1, we set out to characterise it at pH's 5.0 and 8.0 (Cabezon *et al.*, 2000b). Initially we measured density increments for IF_1 solutions dialysed to equilibrium in buffers at these pH's and obtained values of ρ_0 = 1.002128 and 1.000957 g/ml and ϕ' = 0.759 and 0.760 ml/g respectively at pH's 5.0 and 8.0, and these values were used for all subsequent calculations.

Sedimentation equilibrium runs, at various loading concentrations and these pH's, gave plots of $\overline{M}_{w,app}$ against concentration for wild type protein as shown (Fig. 3, upper panels). The monomer molecular mass is 9580 Da, so the wild type protein is predominantly dimeric at pH 5.0 and tetrameric at pH 8.0. For a better test of possible models, the original optical density against radius data were fitted directly (as described above). In practice, the best fit at pH 5.0 was for a non-ideal "monomer", where in practice the "monomer" was a dimer of the protein. Similarly, the data at pH 8.0 were best fitted by a tetramer, with no non-ideality in this case. The validity of the fits is shown by plots of the residuals between the calculated and

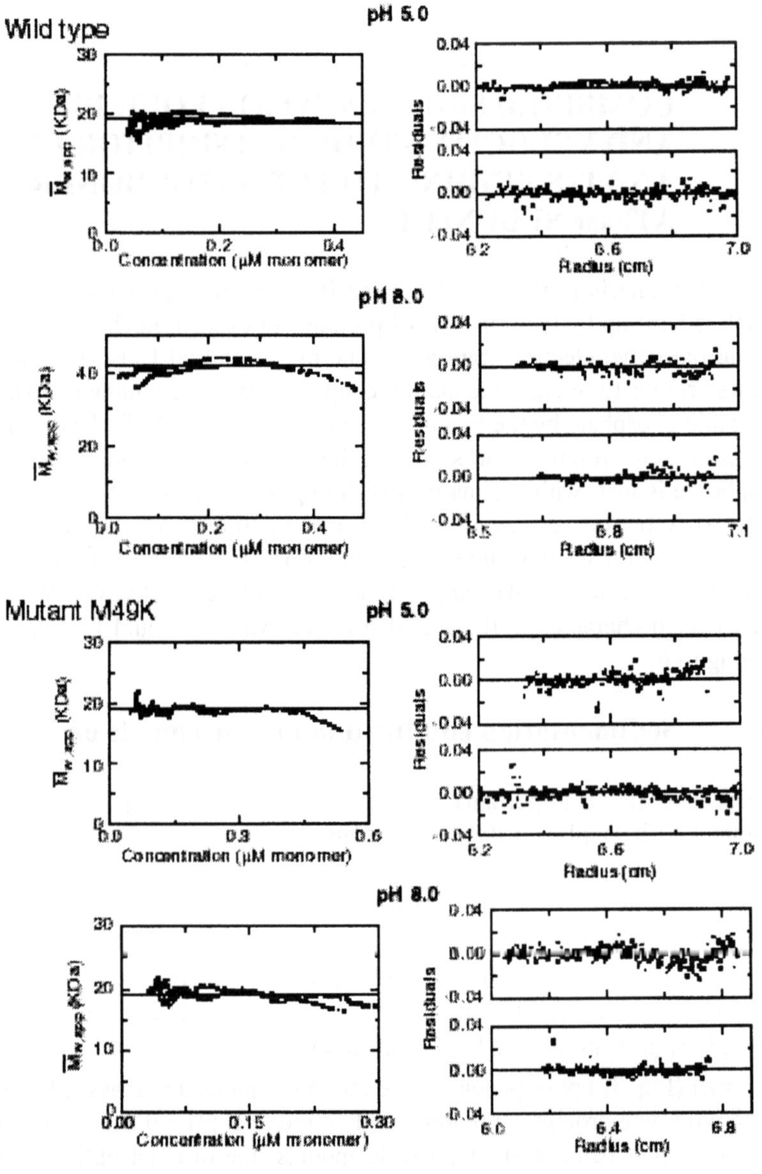

Figure 3. Plots of results from sedimentation equilibrium on IF_1 wild type and mutant M49K, at pH's 5.0 and 8.0. Plots of $\overline{M}_{w,app}$ against concentration are shown for each sample (on the left), with curves fitted from the best-fitting model – in each case a "monomer", with non-ideality for wild type at pH 5.0 but ideal in all other cases. In practice, the "monomer" was dimeric for wild type at pH 5.0 and for mutant M49K at pH's 5.0 or 8.0, but tetrameric for wild type at pH 8.0. Examples of residuals between the calculated and experimental data at each radius point, for fitting of the models, are shown on the right. From Cabezon *et al.* (2000b).

measured absorption at each radial point. These are essentially random and show maxima of ~0.02 D_{280}, while the data goes to ~2.0 D_{280}. The lines in the plot of $\overline{M}_{w,app}$ against concentration come from the fitted model. Attempts to treat either set of runs as a "monomer"/"dimer" equilibrium resulted in K_d falling to the minimum allowed in the fitting, even when this was set below the minimum which could be detected with this optical system.

This experiment strongly suggests that IF_1 is dimeric under the conditions (pH 5.0) where it inhibits the F_1 subunit and that these dimers then dimerise further (to tetramers) when the pH rises to 8.0 and no inhibition is seen. As a further test, we looked at a mutant, M49K, which inhibits the F_1 subunit at both pH's 5.0 and 8.0. Similar plots of the results are shown (Fig. 3, lower panels), and in this case the protein was dimeric over all concentrations at both pH's. In practice, the best fits were given by the model for an ideal "monomer" (actually a dimer), with no evidence for dissociation being evident.

These results show that the active form of IF_1 is a dimer, and that raising the pH causes these to associate into inactive tetramers. One obvious hypothesis is that the association of the dimers might mask the binding site for F_1, thereby preventing inhibition.

4.2 Sedimentation velocity experiments on $F_1 \pm IF_1$

The isolated F_1 subunit of the mitochondrial ATP-synthetase is not sufficiently stable at pH 5.0 to allow sedimentation equilibrium studies under conditions where the IF_1 inhibitor is active. We therefore studied the interaction using sedimentation velocity, with samples of F_1 without and with IF_1 and at pH 5.0 (Cabezon *et al.*, 2000a). The results from one such experiment are shown in Figure 4. The top panels show sequences of scans, spaced equally in time, and it can be seen that the boundary in each experiment is symmetrical, suggestive of a single species. Direct analysis of dc/dt, using the program DCDT+ (Philo, 2000), showed that F_1 alone had $s_{20,w}$ = 12.1 (±0.7) S and the F_1-IF_1 complexes had $s_{20,w}$ = 17.0 (±0.1) S. In order to test whether there might be more than one component in the boundary, plots of g(s*) against s* were made (Stafford III, 1992; Stafford III, 1994). (Where g(s*) is the fraction of material sedimenting between s* and (s*+δs*).) These plots (Fig. 4, middle panels) again show only a single component. As a final test, model fitting of the dc/dt data with models for either a single or two components was carried out. The best fits were given by the single component model, for both data sets, and plots of the fit residuals against radius are shown (Fig. 4, bottom panels).

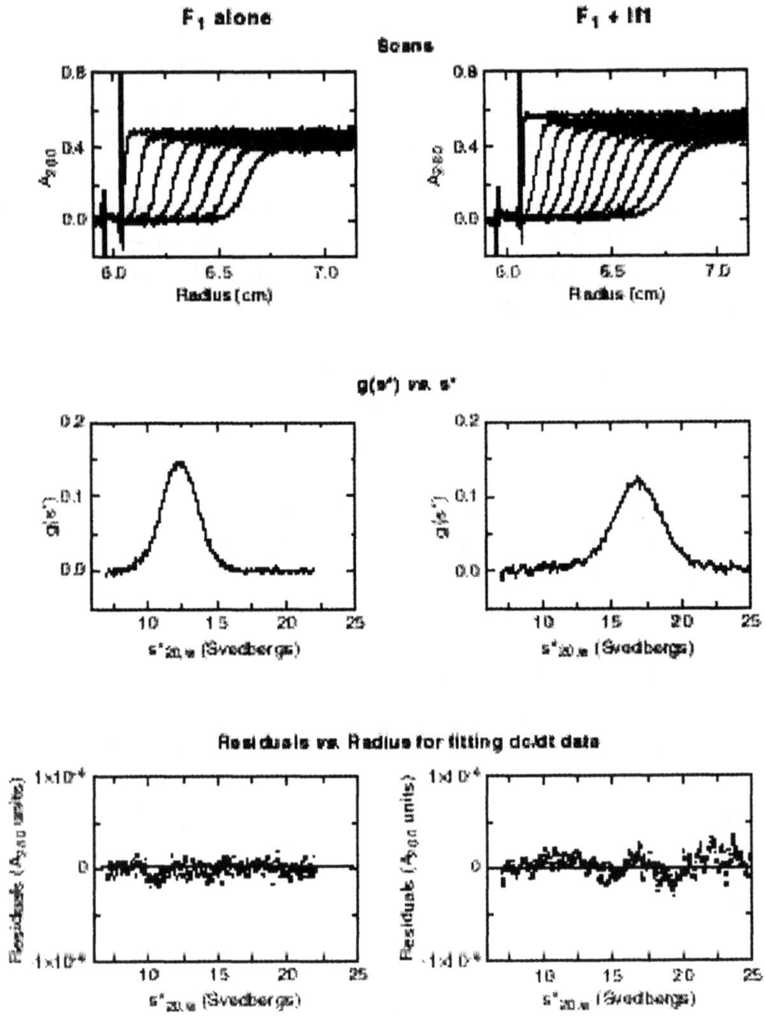

Figure 4. Sedimentation velocity analysis of F_1-ATPase subunit and the purified IF_1-F_1 complex. The data are shown (at equal intervals of time for a given sample) as A_{280} against radius for each sample. Analysis of the data is shown as plots of g(s*) against s* (showing the distribution of the sample with sedimentation coefficient) and also as the residuals (in A_{280} against $s^*_{20,w}$) for fitting a model with a single component, distributed as a Gaussian, to the dc/dt curves. From Cabezon *et al* (2000a).

All of these analyses show that there is predominantly only a single species of F_1 present under each experimental condition, but that this shows an increase in sedimentation coefficient from ~12 S to ~17 S on addition of

IF_1. The mass of the F_1 ATPase subunit (~370 kDa) is so large compared to that of IF_1 (~10 kDa) that free IF_1 will not be visible in the optical system. Nor will the mass difference upon binding of IF_1 to an F_1 subunit be sufficient to alter the sedimentation coefficient measurably. The increase, by a factor of ~1.4, can only be due to dimerisation of F_1 subunits upon binding of the dimeric IF_1.

5. OVERVIEW

I hope that, in this review, I have been able to show you some of the uses of analytical centrifugation in looking at the aggregation state of macromolecules free in solution. By combining equilibrium and velocity measurements, and also measuring density increments so that apparent partial specific volumes can be calculated, one can obtain information about the aggregates which would not be readily available by any other technique. Also, the sensitivity of modern instruments means that only small amounts of material are required even for a series of experiments, and so the methods can be applied to samples which are difficult to obtain in significant quantities – often the experimentally most interesting materials.

REFERENCES

Cabezon, E., Arechaga, I., Butler, P. J. G. & Walker, J. E., 2000a, Dimerization of bovine F_1-ATPase by binding the inhibitor protein, IF_1. *J. Biol. Chem.*, in the press.

Cabezon, E., Butler, P. J. G., Runswick, M. J. & Walker, J. E., 2000b, Modulation of the Oligomerization State of the Bovine F_1-ATPase Inhibitor Protein, IF_1, by pH. *J. Biol. Chem.* **275**: 25460-25464.

Casassa, E. F. & Eisenberg, H., 1964, Thermodynamic analysis of multicomponent solutions. *Adv. Prot. Chem.* **19**: 287-395.

Johnson, M. L. & Straume, M., 1994, Comments on the analysis of sedimentation equilibrium experiments. In *Modern Analytical Ultracentrifugation* (Schuster, T. M. & Laue, T. M., eds.), Birkhäuser, Boston, USA, pp. 37-65.

König, N., Zampighi, G. A. & Butler, P. J. G., 1997, Characterisation of the Major Intrinsic Protein (MIP) from Bovine Lens Fibre Membranes by Electronmicroscopy and Hydrodynamics. *J. Molec. Biol.* **265**:590-602.

Laue, T. M., Shah, B. D., Ridgeway, T. M. & Pelletier, S. L., 992,Computer-aided interpretation of analytical sedimentation data for proteins. In *Analytical Ultracentrifugation in Biochemistry and Polymer Science* (Harding, S. E., Rowe, A. J. & Horton, J. C., eds.), Royal Society of Chemistry, Cambridge, UK., pp. 90-125.

Osmand, A. P., Friedenson, B., Gewurz, H., Painter, R. H., Hofmann, T. & Shelton, E.,1977, haracterisation of C-reactive protein and the complement subcomponent Clt as homologous proteins displaying cyclic pentameric symmetry (pentraxins). *Proc. Nat. Acad. Sci. USA* **74**:739-743.

Pepys, M. B. & Baltz, M. L.,1983,Acute phase proteins with special reference to C-reactive protein and related proteins (pentraxins) and serum amyloid A prtoein. *Adv. Immunol.* **34**: 1-212.

Pepys, M. B., Dash, A. C., Fletcher, T. C., Richardson, N., Munn, E. A. & Feinstein, A.,1978, nalogues in other mammals and in fish of human plasma proteins C-reactive protein and amyloid P component. *Nature (Lond.)* **273**: 168-170.

Perrett, S., Freeman, S. J., Butler, P. J. G. & Fersht, A. R., 1999, Equilibrium folding properties of the yeast prion protein determinant Ure2. *J. Molec. Biol.* **290**: 331-345.

Philo, J. S., 2000, A method for directly fitting the time derivative of sedimentation velocity data and an alternative algorithm for calculating sedimentation coefficient distribution functions. *Analyt. Biochem.* **279**(2): 151-163.

Poget, S. F., Legge, G. B., Proctor, M. R., Butler, P. J. G., Bycroft, M. & Williams, R. L., 1999, The structure of a tunicate C-type lectin from *Polyandrocarpa misakiensis* complexed with D-galactose. *J. Molec. Biol.* **290**: 867-879.

Robey, F. A. & Liu, T.-Y., 1981, Limulin: a C-reactive protein from *Limulus polyphemus*. *J. Biol. Chem.* **256**: 969-975.

Schachman, H. K., 1959, *Ultracentrifugation in Biochemistry*, Academic Press Inc., New York.

Stafford III, W. F., 1992, Boundary analysis in sedimentation transport experiments: a procedure for obtaining sedimentation coefficient distributions using the time derivative of the concentration profile. *Analyt. Biochem.* **203**: 295-301.

Stafford III, W. F., 1994, Methods for obtaining sedimentation coefficient distributions. In *Analytical Ultracentrifugation in Biochemistry and Polymer Science* (Harding, S. E., Rowe, A. J. & Horton, J. C., eds.), The Royal Society of Chemistry, Cambridge, UK.,

Svedberg, T. & Pedersen, K. O., 1940, *The Ultracentrifuge*, Oxford University Press, London and New York. pp. 359-393.

Tennent, G. A., Butler, P. J. G., Hutton, T., Woolfitt, A. R., Harvey, D. J., Rademacher, T. W. & Pepys, M. B., 1993, Molecular characterization of *Limulus polyphemus* C-reactive protein. I. Subunit composition. *Eur. J. Biochem.* **214**: 91-97.

Vermorken, A. J. M., Hilderink, J. M. H. C., Dunia, I., Benedetti, E. L. & Bloemendahl, H., 1977, Changes in membrane protein pattern in relation to lens cell differentiation. *FEBS Lett.* **83**: 301-306.

Mass Spectrometry of Peptides and Proteins -
Principles and Features of Electrospray/Ionization-Mass Spectrometry (ESI-MS) and Matrix-Assisted Laser Desorption/Ionization-Mass Spectrometry (MALDI-MS)

KERSTIN STRUPAT[1] AND WOLFGANG METELMANN[2]
Institute for Medical Physics and Biophysics, [1] Laser Mass Spectrometry Group, [2] Biomedical Analytics Group, Westfälische Wilhelms-Universität Münster Robert-Koch-Str. 31, D-48149 Münster, Germany

1. INTRODUCTION

The ionization techniques Electrospray/Ionization (ESI) and Matrix-Assisted Laser Desorption/Ionization (MALDI) have revolutionized biological mass spectrometry. Their introductions in the late eighties enable for mass analysis of very different kinds of biological molecules such as peptides, proteins, (Yates, 1998) glycoconjugates, DNA and RNA, (Gross and Hillenkamp, 2001) and synthetic polymers (Nielen, 1999) by first, producing ionized gas phase molecular ions of a condensed sample and then separating these ions by their *mass-to-charge- (m/z) ratios* in mass analyzers. In order to allow for a mass-to-charge separation of ions in a mass analyzer, analyte molecules need to be available as ionized, isolated entities in the gas phase. Both ionization techniques enable this in an extremely soft way, meaning that a low fragmentation yield of the analyte molecules is introduced (compared to conventional ionization techniques such as electron impact ionization or chemical ionization). Inherent (more related to MALDI) or induced fragmentation (MALDI and ESI) can be used for sequencing of analyte molecules.

Considerable information can be taken from mass spectra of biological samples. (Gaskell, 1997; Gross and Strupat, 1998) Besides the determination

Supramolecular Structure and Function 7, edited by Pifat-Mrzljak.
Kluwer Academic/Plenum Publishers, New York, 2001.

of the *molecular mass* M of a given compound (often named *molecular weight* M_r), the identification of proteins by mass determination of their proteolytic fragments is a very often used tool. Mass spectrometry is also capable of providing structural, i.e. sequence information of peptides. Together with enzymatic degradation, the carbohydrate content of a glycoprotein can be evaluated at least semi-quantitatively. Furthermore, the quaternary structure of protein complexes, and the interaction between proteins and ligands or metal ions can be studied by mass spectrometry.(Loo, 1997; Smith *et al.*, 1997) In the rapidly expanding field of *proteomics* high quality mass data – namely accurate mass determination – represents the key to unambiguous protein identification by peptide mass mapping and to determination of post-translational modifications of proteins grown under different experimental conditions. The accuracy of mass determination in peptide mass analysis obtained by ESI- and MALDI-MS has increased from less than 500 ppm (0.5 u at 1000 u) ten years ago to 10 – 50 ppm (0.01u – 0.05 u at 1000 u) at present. The striking feature of an improved mass accuracy is the dramatic reduction of false hits in data base interrogation for protein identity.(Jensen *et al.*, 1996, Shevchenko *et al.*, 1996, Clauser *et al.*, 1999)

It should be mentioned in this context that the expressions *accuracy of mass determination* and *precision of mass determination* are important values to qualify the information derived from mass spectra. While the precision of mass determination is related to the reproducibility of mass measurements), the accuracy of mass determination describes how accurate the mass of a peptide or protein can be measured. The accuracy of mass determination therefore expresses the deviation between the determined determination (mean value $M_{average}$ and standard deviation σ of a given set of mean value $M_{average}$ and the theoretical molecular mass M of the given analyte molecule.

Mass resolution R is another important expression to be mentioned: For time-of-flight mass analyzers the so-called *peak width definition* is a common standard.(Price, 1991) It considers a single peak in a mass spectrum made up of singly charged ions at mass M. The mass resolution R is expressed as $M/\Delta M$ where ΔM is the width of the peak at a height that is a specific fraction of the maximum peak height. A common standard is the definition of resolution R based upon the full peak width ΔM at half maximum (FWHM). Considering a signal at mass M = 1000 u and a peak width of this signal of $\Delta M = 0.5$ u (FWHM), mass resolution is $R = M/\Delta M = 2000$. Different definitions for other mass analyzers are typically taken also into account but are not mentioned here. High mass resolution of a mass analyzer is neither a necessary prerequisite for accurate mass determination nor a sufficient one, since accurate mass determination depends also upon well-defined calibration standards of known mass, correct calibration procedures, and precise (i.e. reproducible) peak shapes. However, high mass resolution helps to obtain high mass accuracy, if the latter conditions are optimized and adapted to the analytical problem. As a rough guideline: if mass resolution is R = 10000 at mass M = 1000 u, i.e. the peak width ΔM = 0.1 u (FWHM), the achievable accuracy of mass determination is about or better than one tenth of the peak width ΔM. This means that the mass of an ion at mass 1000 u can be determined with an accuracy of at least 0.01 u at 1000 u (10 ppm).

A state of the art mass spectrometer is able to resolve adjacent isotopes of peptides in the mass range up to M = 5000 u. In other words, the molecular ion containing only ^{12}C-isotopes can be distinguished from the molecular ion containing one ^{13}C-isotope. (compare to Figures 3 and 5)

A *mass spectrometer* is built up by mainly three parts: an *ion source* for the production of ionized isolated entities in the gas phase, a *mass analyzer* for mass-to-charge separation of the ions produced in the ion source, and a *detection system* (typically realized by a secondary electron multiplier or a multi channel plate). The detector is followed by a suitable data processing system including a PC. Mass analyzers and detection systems have to be under high vacuum (10^{-6} mbar or lower), in order to minimize number of collisions with residue gas molecules; this enables a long mean free path of the ions in the mass spectrometer.

MALDI and ESI are two different ionization methods and therefore require a different ion source design. Although this aspect will not be discussed in detail here, the principle features and characteristics of both ionization techniques will be discussed further down. Different kinds of mass analyzers allow to separate MALDI- or ESI-produced ions by their m/z ratios. In this paper the principle of time-of-flight (TOF) mass analyzers and of quadrupole mass filters will be discussed briefly, because those are currently most often used for m/z-separation of MALDI- and ESI-produced ions. The principle set-up of these separating devices is shown in Figure 1.

1.1 Time-of-flight mass analyzers (TOF)

The basis of mass-to-charge separation in a time-of-flight mass analyzer lies in the determination of the flight time of ions of different m/z-ratios. The time the ions need to pass the distance **s** of the so-called field free drift region after being accelerated to the kinetic energy $E_{kin} = \frac{1}{2} \cdot m \cdot v^2$ is determined by the accelerating potential U_{acc} in the ion source (see Figure 1 a). Taking into account a singly or multiply charged ion (elementary charge e, number of charges z) means that the kinetic energy equals $z \cdot e \cdot U_{acc}$:

$$E_{kin} = \frac{1}{2} \cdot m \cdot v^2 = z \cdot e \cdot U_{acc} \ . \qquad (1)$$

By this relationship the velocity **v** of an ion after reaching the end of the acceleration region in the ion source is given by

$$v \ = (2 \cdot z \cdot e \cdot U_{acc} / m)^{1/2} . \qquad (2)$$

In the field free drift region of length **s** the motion of the ions is non-accelerated and all the ions move with the velocity **v** they have obtained by the acceleration in the ion source, this means $v = s / t$ and therefore

$$s / t = (2 \cdot z \cdot e \cdot U_{acc} / m)^{1/2} . \qquad (3)$$

Rearranging this equation to the time of flight t of the ions becomes

$$t \ = s \cdot (m / 2 \cdot z \cdot e \cdot U_{acc})^{1/2} . \qquad (4)$$

From this equation it becomes obvious that the time of flight of an ion is proportional to the square root of its mass-to charge ratio m/z:

$$t \ \propto (m / z)^{1/2} . \qquad (5)$$

This relationship is the underlying principle for mass separation by time-of-flight measurement. To fill the relationship with words, one can say that ions of small m/z-ratio travel faster than ions of large m/z ratio after being accelerated in the same field in the ion source. For MALDI-produced ions this relation is even simpler to understand, since MALDI produces predominantly singly charged ions (see further down): Ions with small molecular mass (low m/z) have short flight times, while heavy ions (high m/z) have long flight times (see Figure 1a). For MALDI-ion time of flight times are – depending on the instrumental geometry and acceleration potential in the µs-range. Flight time measurements can be performed in e.g. so-called linear TOF systems as shown in Figure 1a.

A so-called reflector TOF-system is also realized in which the field free drift region is divided in two parts and separated by an electrostatic ion mirror. This helps to improve mass resolution R as described elsewhere, but is not discussed further here.(Hillenkamp *et al.*, 1991)

TOF-systems are most frequently used for MALDI-MS applications, but *hybrid instruments* (quadrupole mass filter combined with a TOF mass analyzer) become more and more important.

1.2 Quadrupole mass filter

A quadrupole mass filter is constructed by two pairs of bars in a cylindrical symmetrical arrangement (Figure 1a). Ideally, the bars have an hyperbolic cross section, but typically a circular cross section is realized for the ease of construction. A potential of a superposition of a DC- and an AC-voltage is applied to each pair of bars. With a DC-voltage U and an AC-voltage $V \cdot \cos \omega t$ the potential applied is $\pm \Phi/2 = U + V \cdot \cos \omega t$. The field built up in the space between the bars by this potential is a so-called *quadrupole field* with the feature of a remaining force directed towards the center of the field; this aspect is particularly interesting because it enables for the focussing of ions (see further down). Hyperbolic shaped equipotential lines (values of equal electric field) are characteristic for the quadrupole field (Figure 1c).

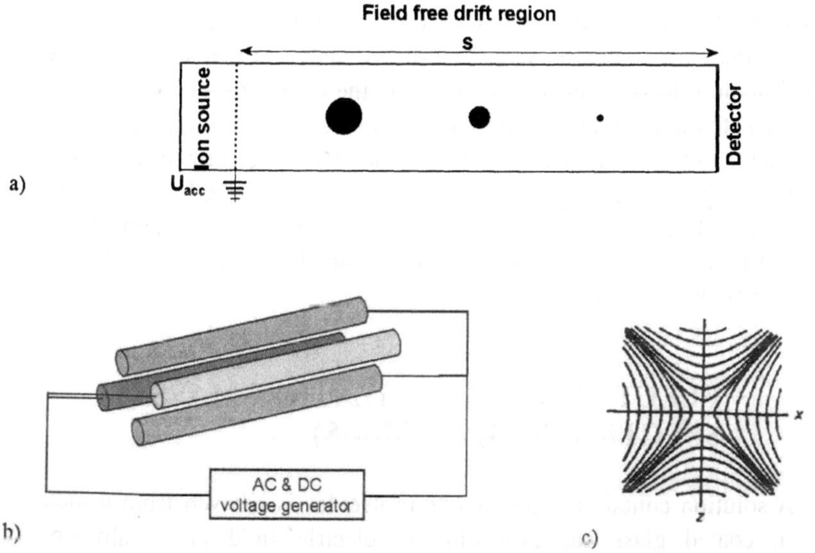

Figure 1: a) schematic set-up of a linear time-of-flight (TOF) mass analyzer. Different mass-to-charge ratios are separated by different arrival times, tof $\propto (m/z)^{1/2}$, b) schematic set-up of a quadrupole mass filter. Different mass-to-charge ratios are brought to stable motions through the quadrupole field by simultaneous change of the AC- and DC-voltages (scanning of U/V-ratio), c) hyperbolic shaped equipotential lines of the quadrupole field, representing values of equal electric field strength.

Ions are brought into the space between the bars with a non-zero velocity component v_y parallel to the bars (Figure 1b). The proper choice and the variation of the amplitudes U and V enable for the transmission of ions of different m/z-ratios through the quadrupole field. At a given U/V-ratio only a certain m/z-range can be transmitted (stable motion in x- and z-direction), while the other m/z-ratios hit the electrodes (unstable motion in x- and/or z-direction, their amplitudes grow exponentially). Scanning the U/V-ratio allows for the accumulation of a whole mass spectrum. This is why the quadrupole mass filter is a *scanning device* and in contrast to the TOF-mass analyzer different m/z-ratios cannot be detected simultaneously.

The motion of an ion of a given m/z-ratio might be understood as a subsequent focussing to and defocussing from the center of the quadrupole. According to Newton's law $F = m \cdot a$, the equations of motions of an ion enable for the description of any trajectory of an ion at any time in the quadrupole field. The equations of motion of an ion in the quadrupole field are described in certain differential equations (*Matthieu Equations*); this is described elsewhere in detail.(Paul 1990)

The interest in quadrupole mass filters is still high and quadrupoles or triple quadrupole mass filters were the working horses for many years in ESI-MS. Today the interest has moved to *hybrid instruments* (quadrupole mass filters are combined with TOF-mass analyzers). In such a device the quadrupole selects a desired m/z-ratio by the choice of a certain U/V-ratio. The selected ion is transported to a collision cell in which the selected ion is fragmented by *collision induced dissociation* (CID, performed with Argon as a collision gas). The fragments are mass analyzed by a TOF-system which is adjacent and perpendicular to the quadrupole arrangement.(Chernushevich et al 1999) Such a device enables MS/MS-application (sequencing of peptides, glycoconjugates, etc.)

2. ELECTROSPRAY/IONIZATION-MASS SPECTROMETRY (ESI-MS)

A solution containing the analyte molecules is sprayed from a metal or metal coated glass capillary into an electric field under atmospheric conditions. The electric field generates a mist of highly charged droplets containing analyte molecules. The highly charged droplets move down a potential gradient ΔU and a pressure gradient Δp – both desolvate or liberate the analyte molecule from solvent molecules - towards the mass analyzer. A schematic drawing of this is given in Figure 2.

Depending on the solution conditions, basic amino acids are protonated or acidic amino acids are deprotonated in solution. Three major steps are necessary for the production of gas phase ions from electrolyte ions in solution by electrospray: a) production of charged droplets at the electrospray capillary tip, b) shrinkage of the charged droplets by solvent evaporation and droplet des-integration, leading to very small highly charged droplets capable of producing gas phase ions, and c) the actual mechanism by which gas phase ions are produced from the very small and highly charged droplets. A comprehensive understanding of the ESI-process is given by Kebarle and Tang (1993); the paper takes into account the model proposed by Iribarne and Thomson (1976) (*ion evaporation model*) and the model proposed by Dole (1968) (*charged residue model*).

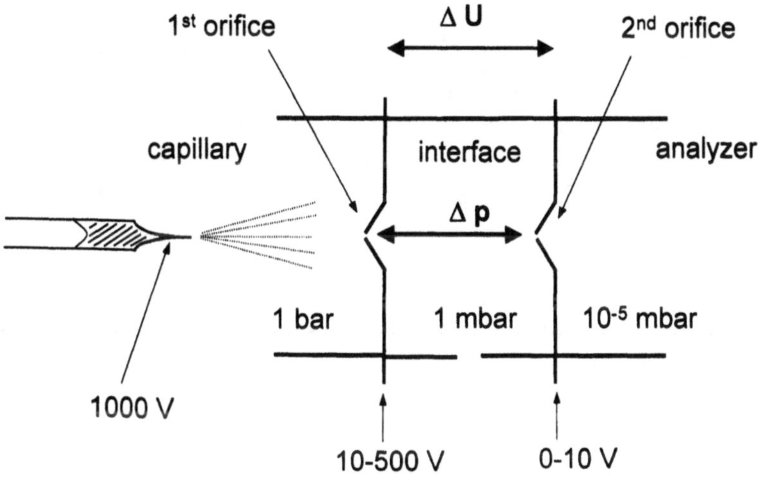

Figure 2: Electrospray/Ionization. A solution containing the analyte molecules is sprayed into an electric field producing a mist of charged droplets. A potential gradient ΔU and a pressure gradient Δp enable the complete desolvation of finally multiply charged analyte molecules in the gas phase.

2.1 ESI-sample preparation

Samples are typically dissolved in solvent mixtures of water/acetonitrile or water/methanol (50/50, v/v) containing 1% formic acid or acetic acid and sprayed from an electrically contacted capillary (metal capillary (Wilm and Mann, 1994) or metal coated glass capillary (Wilm and Mann, 1996)) into an electric field ($E_c \approx 10^6$ V/m) under atmospheric conditions. This spray produces charged solvent droplets containing analyte molecules. A liberation of charged analyte molecules from the solvent enables a subsequent mass analysis of the compounds in suitable mass analyzers such as quadrupole mass filters, ion traps, or time-of-flight mass analyzers.

2.2 Appearance of ESI-mass spectra

ESI-MS generates highly charged ions of peptides or proteins. An ESI-mass spectrum is characterized by a number of signals each differing by one charge. Such a distribution of charge states is typically produced by multiple protonation (positive ion mode) or deprotonation (negative ion mode) of the species, but cation formation is also well known to occur. An ion signal of a species of mass M produced by n-fold protonation (mass of a proton m_H) has the mass $M + n \cdot m_H$; these signals are assigned as $(M + n\ H)^{n+}$ or M^{n+} (positive ion mode) in the mass spectrum. The signals occur at mass-to-charge ratios $m/z = (M + n \cdot m_H) / n$. A mass range of the mass analyzer of up to m/z 2000 is therefore sufficient for many ESI-applications.

Dissolving peptides and proteins under denaturing conditions (organic solvent plus water containing acid) as described above, results in more or less pronounced unfolding of the analyte molecule in solution, therefore increasing the number of protonation sites. Acetic or formic acid in the solvent then ensure protonation of these basic amino acids. As a rough rule, the number of basic amino acids determines the maximum number of charges z observed in positive ESI-MS. Peptides are typically observed with 2 to 5 protons, $(M+2H)^{2+}$, ..., $(M+5H)^{5+}$, in the mass spectrum depending on the mass of the peptide and the number of its basic residues. The m/z-ratios are typically between m/z = 300 and m/z = 1500.

ESI-MS data of an equimolar mixture of angiotensin I, substance P, and neurotensin (1 pmol/µl each peptide, all peptides of human origin) are shown in Figure 3. The triply charged species $(M+3H)^{3+}$ are observed predominantly for each of these three peptides, accompanied by the doubly charged species; angiotensin I is also observed as fourfold charged species (m/z = 325). The charge state information observed for the three peptides, therefore, is in good agreement with the number of basic amino acids plus one charge for the N-terminus. Angiotensin I has four likely protonation sites, while substance P and neurotensin have three potential protonation sites (i.e. the sum of basic amino acid residues (Lys, Arg, His) plus one protonation site at the amino terminus). Mass resolution R = 6000 (FWHM) is obtained for each peptide shown in Figure 3 (see insets). If the triply charged species of angiotensin I and neurotensin are used for calibration of the mass scale $300 \leq m/z \leq 1000$, the mass of substance P is determined with an accuracy of $\approx$ 10 ppm (Figure 3, internal calibration, i.e. using signals of the same spectrum to calibrate the mass scale). The spacing between ^{12}C

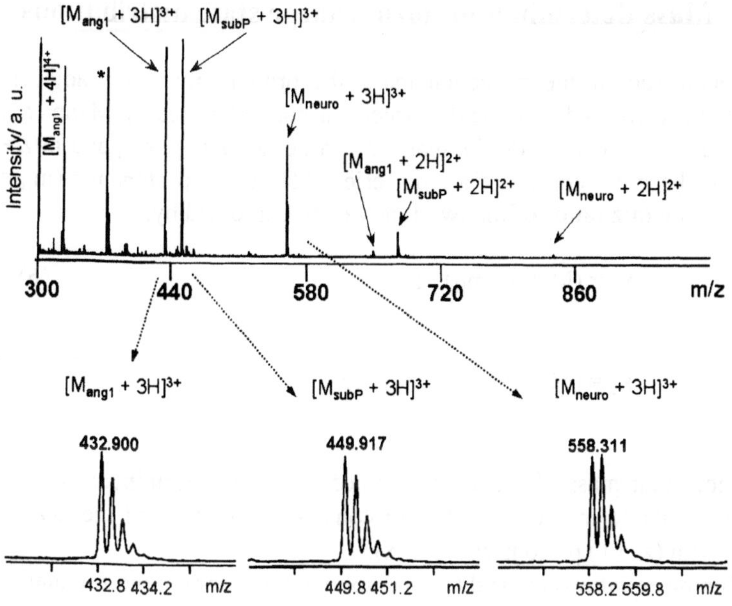

Figure 3: ESI-mass spectrum of an equimolar mixture of the three peptides angiotensin I (human), substance P (human), and neurotensin (human) sprayed using denaturing solvents. Each analyte molecule has a concentration of 10^{-6} M. The insets show the region of the triply protonated signals $[M + 3H]^{3+}$ of the peptides. The signal marked by an asterisk * is due to an impurity.

isotopic peak (mono isotopic signal) and the peak including one ^{13}C isotope is equal to 1/(charge state z). In the example shown in Figure 3, the spacing between adjacent signals of angiotensin equals to m/z = $0.\bar{3}$ and therefore corresponds to charge state z = 3. Knowing the charge state of a given peptide signal enables the determination of the mono isotopic mass of a peptide M (see further down).

For proteins, a broader charge state distribution and more highly charged species are observed. As a rough guide, one charge per 1000 u is observed using denaturing solvent conditions and the mass-to-charge range covered by the distribution is between m/z = 500 and m/z = 2000. Often the width of this distribution of charge states is about half that of the highest charge state.(Smith *et al.*, 1990) For a protein such as myoglobin (apo-form, horse, 16951.49 u) (Zaia *et al.*, 1992) signals will appear between m/z = 700 and m/z = 1700 with 24 (m/z = 707.3, $[M+24H]^{24+}$) to 10 (m/z = 1696.15, $[M+10H]^{12+}$) charges (Figure 4 a).

2.3 Mass determination form charge state distributions

For reconstructing the molecular mass of a protein any two of adjacent peaks are sufficient to determine the molecular mass of a species.(Mann *et al* 1989) Referring to Figure 4a, the mass-to-charge ratios m/z = x_1 and m/z = x_2 are two adjacent members of an ion series obtained by protonation (m_H = 1.00794 u); the m/z ratios of the two ions can be expressed by:

$$x_1 = (M + z \cdot m_H) / z \tag{6}$$

$$x_2 = (M + (z + 1) \cdot m_H) / (z + 1) \tag{7}$$

with

M : molecular mass of the analyte molecule, to be determined; x_1 : m/z ratio of the analyte molecule with z protons; x_2 : m/z ratio of the analyte molecule with (z + 1) protons and $x_1 > x_2$.

Combining equations (6) and (7) and solving them for charge z, equation (8) allows the determination of the charge state z of the ion signal at m/z = x_1:

$$z = \frac{x_2 - m_H}{x_1 - x_2} \; . \tag{8}$$

The number of charges z of the ion signal of m/z = x_1 allows the determination of the molecular mass M of the analyte molecule :

$$M = z \cdot x_1 - z \cdot m_H \;= (z + 1) \cdot x_2 - (z + 1) \cdot m_H \; . \tag{9}$$

Taking the spectrum of Figure 4a as an example, two adjacent charge states might be considered: x_1 = 998.2 and x_2 = 942.8, the charge state of x_1 is calculated to z = 17 from these numbers. The molecular mass of apo-myoglobin (16951.49 u) can be determined with a precision of better than ± 50 ppm (± 0.85 u) using external mass calibration.

2.4 Electrospray/ionization-mass spectrometry of non-covalent complexes

Non-covalent complexes play an important role in living matter since the binding and release of molecules to and from each other is essential in control and regulation mechanisms in cells. ESI-MS is a powerful tool for the analysis of non-covalent complexes such as protein-protein-, protein-ligand (cofactor, receptor, inhibitor)-, or protein-DNA-complexes. The main challenge in mass analyzing biological non-covalent complexes by ESI is to enable preparations and ionization conditions which preserve the non-covalent complex and which reflect biological reality.

There are several pitfalls for a non-covalent complex to fall apart, either in solution, in the spraying process, or in the ESI-interface due to the pressure gradient from 1 bar at the capillary tip to 10^{-5} mbar or even lower in the region towards the mass analyzer (compare to Figure 2). In order to ensure a complex in solution, native solution conditions (buffered, aqueous solutions) are chosen. Furthermore, to keep the complex in its native state during the ionization process requires a careful control of different parameters of the ESI-process and the ESI-interface.

In contrast to apo-myoglobin measured from denaturing solvents (which produces a wide charge state distribution, Figure 4a), myoglobin measured from native solvents shows a narrow charge state distribution (Figure 4b). The dissolvation of myoglobin in e.g. 10 mM ammonium acetate (water as solvent, pH 7.5) enables for the observation of the non-covalent complex between protein chain (apo-myoglobin) and its prosthetic heme group (Figure 4b). The complex between apo-myoglobin and its prosthetic heme group can be, however, dissociated in the ESI-interface by *collision induced dissociation* (CID) with residual gas molecules, depending on the interface conditions(Figure 4 c, compare also Figure 2). Simply, increasing the potential gradient dU the complex dissociates and the loss of the heme group from myoglobin is observed. This is reflected by both the actual m/z-ratio and the thereby determined mass M of the molecule (see Figures 4b and 4c).

3. MATRIX-ASSISTED LASER DESORPTION/ IONIZATION-MASS SPECTROMETRY (MALDI-MS)

A mixture (sample) containing the analyte and a so-called matrix is produced on a metallic support (target). A UV- or IR-laser beam of some nanosecond pulse duration hits the sample (in vacuum) and desintegrates the material. Neutrals and (luckily) ions are produced of both matrix and analyte molecules. MALDI-produced ions are (most often) mass analyzed in a TOF-mass analyzer, since it is easy to trigger (i.e. switch on) the time-of-flight measurement by a short, well defined laser pulse. For this purpose a part of the laser beam is deflected to a photodiode and the signal produced by the photodiode sets the time zero signal for the time-of flight measurement.

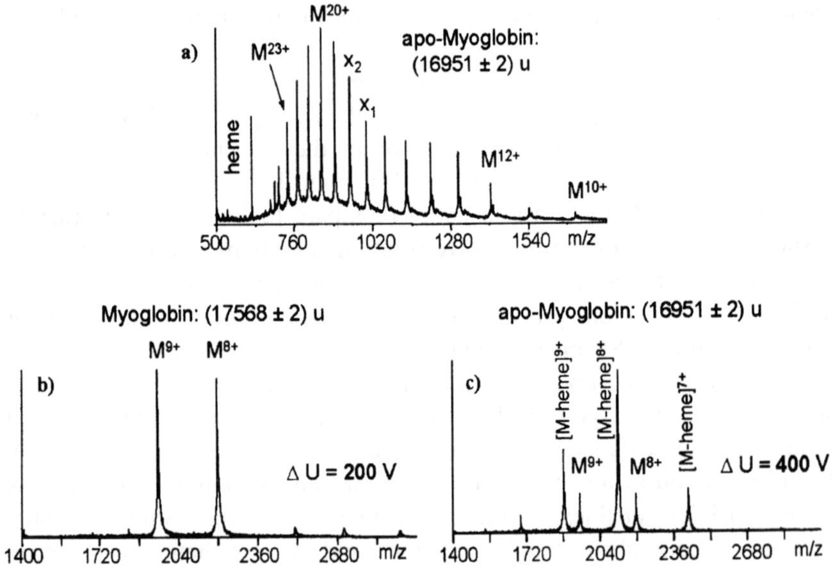

Figure 4: ESI-mass spectra obtained from different solution conditions. myoglobin (horse heart) as an analyte molecule. a) myoglobin dissolved in acetonitrile/H_20/formic acid (49.5/49.5/1, v/v/v). A series of mulitiply charged ions is obtained from z = 10 (m/z = 1696.2) to z = 24 (m/z = 707.3). The non-covalent complex between protein (apo-myoglobin) and its prosthetic heme group dissociates in solution. The mass of apo-myoglobin can be determined from this ion series.(see text); b) myoglobin dissolved in 10 mM ammonium acetate (pure water as solvent, pH 7.5). A narrow charge state distribution (z = 8, m/z = 2197, and z = 9, m/z = 1953) is obtained which is indicative of the much closer structure of the protein under native solution conditions. Under mild interface conditions (ΔU = 200 V) the complex between apo-myoglobin and its prosthetic group survives the transfer from the liquid to the gas phase; c) Same sample measured under less soft interface conditions (ΔU = 400 V). A collision of myoglobin with residual gas molecules is more efficient now, since myoglobin gains more kinetic energy between subsequent collisions. The non-covalent complex does no longer survive the transfer from the liquid to the gas phase.

3.1 The idea of the matrix

The general idea of a MALDI-sample preparation is to mix a solution of analyte molecules of relatively low concentration with a so-called *matrix* solution of relatively high concentration. The matrix - typically a small organic molecule - has mainly three tasks which finally enable a desorption/ionization of analyte molecules independent of its individual features. The first is to strongly and resonantly absorb the irradiated laser wavelength which requires an aromatic system as a matrix compound is in the case of ultraviolet-MALDI (UV-MALDI, electronic excitation) or an aromatic or alliphatic system in the case of infrared-MALDI (IR-MALDI, rotational-vibration excitation). Specific matrix compounds can be derivatized by various functional groups (-OH, $-NH_2$, OCH_3, ...) to accommodate the absorption of the matrix to the irradiating laser wavelength. The second task of the matrix is to force separation of analyte molecules from each other (*matrix isolation*): The required analyte separation (*matrix isolation*) is achieved by a large molar excess of the matrix resulting in a molar matrix-to-analyte ratio between 10^3 and 10^6. Therefore, the matrix concentration is $\approx 10^{-1}$ mol/l, while the analyte concentration is in the range between 10^{-4} mol/l and 10^{-7} mol/l depending on the size of the analyte molecule and its purity. A peptide or protein concentration of 0.1 g/l (10^{-4} mol/l for mass 1 000 u and 10^{-6} mol/l for mass 100 000 u) is therefore sufficient for a successful MALDI-mass analysis. The third task is understood in its ability to help or to initiate analyte ionization: (Karas *et al.*, 1985, Karas and Hillenkamp, 1988, Karas *et al.*, 1989) It is assumed that the choice of the matrix plays an important role in analyte ionization: a proton transfer from the electronically excited matrix compound to the analyte molecules in the expanding cloud of ions and neutrals (plume) desorbed by the laser pulse (some nanoseconds) might be responsible for ionization. A model taking into account the different physical properties of UV- and IR-MALDI has yet to be explored.(Ehring *et al.*, 1992, Zenobi and Knochenmuss, 1999, Karas *et al.*, 2000)

3.2 MALDI-sample preparation

Two different main sample preparation techniques should be distinguished: fast evaporation technique and dried droplet preparation. The *fast evaporation* or *thin layer preparation* introduced by Vorm and Mann (1994), Vorm *et al.* (1994) results in a very homogeneous sample morphology. It is suited to matrix compounds which are almost water-insoluble such as α-cyano-4-hydroxy cinnamic acid (ACCA).(Beavis *et al.*, 1992) The MALDI-sample is prepared by first producing a thin matrix layer on the target and afterwards spotting the analyte solution on top of this matrix layer. For this purpose, a saturated matrix solution dissolved in acetone is spread over the target, the solvent evaporates quickly, leaving a thin, dry matrix layer behind. The analyte solution (preferentially in a slightly acidic solution, containing 0.1% TFA) is added on top of the matrix layer which cannot be dissolved completely by the solvent. Whether this results in analyte incorporation into the ACCA-matrix crystals or an attachment of analyte molecules to the matrix surface is still debated.(Horneffer *et al.*, 1999) The thin layer preparation technique using ACCA-matrix is preferentially used for peptide mass mapping with the aim of protein identification.

The *dried droplet preparation* is best suited for water-soluble matrix compounds such as 2,5-dihydroxy-benzoic acid (2,5-DHB) or mixtures with 2-hydroxy-5-methoxy-benzoic acid (DHBs), (Karas *et al.*, 1993) 3-hydroxy picolinic acid, (both UV-MALDI) or succinic acid (IR-MALDI) and results in a more heterogeneous sample morphology. The MALDI-sample is prepared by mixing analyte and matrix solution directly on the target and air-drying the sample. Co-crystallisation of matrix and analyte results in matrix crystals with dimensions of the 100 μm range or larger. Several helpful hints about sample preparation, including sample purification steps on a μl-scale prior to matrix incubation and on-target reactions such as DTT-reduction, are described in the literature.(Kussmann *et al.*, 1997, Gobom *et al.*, 1999)

3.3 Appearance of MALDI-mass spectra

When accumulating spectra in the positive ion mode, single protonation of peptide and protein species is the most frequent mechanism. The most abundant ion signal is, therefore, assigned as $[M+H]^+$ or M^+. Depending on sample purity, more or less pronounced sodium- and potassium adduct ions $[M+Na]^+$ or $[M-H+Na+K]^+$ occur which, in the case of poor resolution, broaden the signal to the higher mass side. This becomes a severe problem with increasing mass of the protein, because cation adducts cannot be resolved anymore from the protonated species. The singly charged (protonated) ion species M^+ is accompanied by less intense doubly and more highly charged ions (M^{n+}) and some still less abundant singly or more highly charged (most likely) gas phase-induced oligomers ($m\,M^{n+}$) of the species. The fact that MALDI produces predominantly singly charged ions significantly facilitates mass spectra interpretation compared to ESI, if more than one species are present in the sample. Even in the analysis of mixtures, signals corresponding to the same analyte molecule are easy to assign, making MALDI-mass analysis most straightforward for peptide mass mapping.(Jensen *et al.*, 1996, Shevchenko *et al.*, 1996, Clauser *et al.*, 1999)

In peptide mass analysis and positive ion mode, signals corresponding to singly charged protonated species $(M+H)^+$ are almost the only signals obtained. Figure 5 shows the same peptide mixture as shown in Figure 3, but now using ACCA-matrix prepared as a thin layer analyzed by MALDI-MS; angiotensin I, substance P, and neurotensin are added on top of the dried matrix layer. Each peptide has a concentration of 1 μM. Interestingly, no signals corresponding to dimers ($2M^+$) and no doubly charged species (M^{2+}) are observed. This is very typical for the mass analysis of peptides by MALDI-MS.

With increasing mass, more highly charged signals such as $[M+2H]^{2+}$, $[M+3H]^{3+}$, and singly or doubly charged oligomers $[2M+H]^+$ or

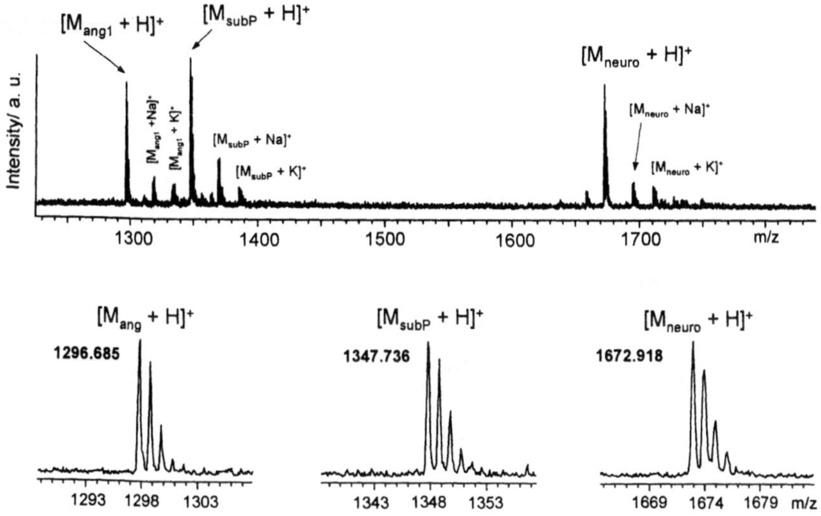

Figure 5: MALDI-mass spectrum of an equimolar mixture of the three peptides Angiotensin 1 (human), Substance P (human), and Neurotensin (human) using ACCA-matrix and fast evaporation technique for sample preparation. Each analyte molecule has a concentration of 10^{-6} M. Only singly charged species appear in the mass spectrum (as protonated species and (less abundant) as sodium and potassium attached species). As insets the singly protonated species are shown.

$[3M+2H]^{2+}$ may show up in the mass spectrum. As an example of a medium sized protein Figure 6 shows apo-myoglobin (5 μM) prepared with DHBs-matrix.

4. PROTEIN IDENTIFICATION BY PEPTIDE MAPPING AND PEPTIDE SEQUENCING BY MEANS OF MASS SPECTROMETRY

The general strategy to identify a protein in a data base is described briefly in the following (for further detail see the paper by Yates, 1998)) First of all, the protein of interest is isolated from a typically crude mixture containing many other proteins, peptides and other biological components. The last step in such a purification/isolation procedure is typically a polyacrylamide gel electrophoresis (PAGE). A gel band (SDS-PAGE) or gel spot (2D-PAGE) is cut with a razor blade, washed and purified from SDS-molecules. The protein in the gel piece is then treated with an enzyme (typically trypsin) in order to produce proteolytic fragments. Those

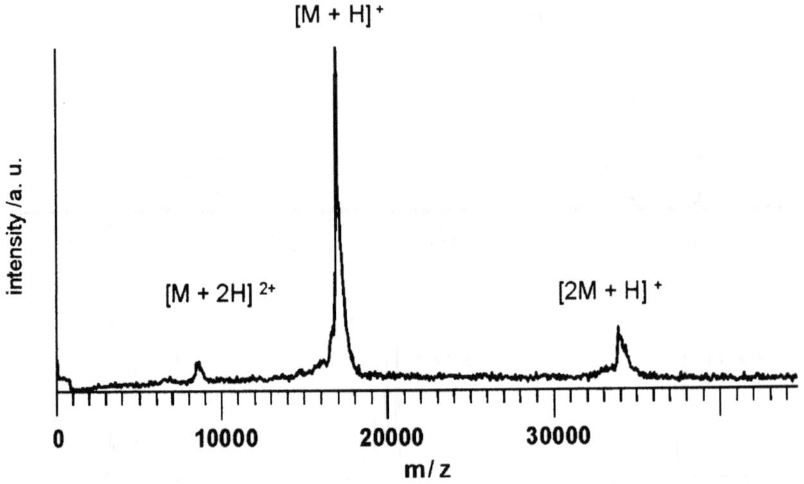

Figure 6: MALDI-mass spectrum of apo-myoglobin. The singly charged species - $[M+H]^+$ - is the most intense ion signal, accompanied by a doubly and triply charged species and a singly charged (MALDI-induced) dimer $[2M+H]^+$.

fragments can be eluted from the gel piece and mass analyzed – after some suitable purification steps – by MALDI-MS. (Shevchenko *et al.*, 1996) MALDI-MS delivers a so-called *peptide mass map*. Search algorithms are available in the internet which enable to interrogate for the identity of the protein by typing in the masses found by the peptide mass map. It should be kept in mind that the more accurate the fragments of the peptide mass map can be determined, the higher the confidence in a subsequent data base search for protein identity.

If still the identity of the protein is ambiguous, the mixture of proteolytic fragments can be analyzed by ESI-MS: The above mentioned hybrid instruments enable to select one of the fragments in a quadrupole mass filter. A subsequent collision induced dissociation in the collision cell specifically cleaves the selected peptide. The cleavage products are mass analyzed in the TOF-mass analyzer. Typically, fragments which derive from a cleavage either in, before or after the peptide bond are observed and allow for reading the peptide sequence.(Yates *et al.*, 1998) The sequence of a given proteolytic fragment makes data base searching for protein identity very reliable. Also post-translational modifications can be read by this sequencing approach. *De novo* sequencing can be performed by this MS/MS-strategy instead of performing conventional Edman-sequencing (Shevchenko *et al.*, 1997).

ACKNOWLEDGMENTS

Data shown in Figures 3 and 4 were obtained using an ESI-oa-TOF-mass analyzer (Mariner, Applied Biosystems, Framingham, MA, U.S.A.) equipped with a nano-spray source. The instrument was kindly given to KS for research work; this support by Applied Biosystems is highly appreciated.

REFERENCES

Beavis, R.C., Chaudhary, T., Chait, B.T., 1992, *Org.Mass Spectrom.* **27**: 156.

Chernushevich, I.V.; Ens, W.; Standing, K.G., 1999, *Anal. Chem.* **71**: 452A.

Clauser, K. R.; Baker, P.; Burlingame, A. L., 1999, *Anal. Chem.* **71**: 2871.

Dole, M.; Mack, L.L. Hines, R.L.; Mobley, R.C.; Ferguson, L.D.; Alice, M.B., 1968, *J. Chem. Phys.* 49: 2240.

Ehring, H., Karas, M., Hillenkamp, F., 1992, *Org.Mass Spectrom.* **27**: 472.

Gaskell, S.J., 1997, *J. Mass Spectrometry* **32**, 677;

Gross, J., Strupat, K., 1998, *Trends in Anal. Chem.* **17**: 470.

Gobom, J., Nordhoff, E.; Mirgorodskaya, E.; Ekman, R.; Roepstorff, P.; 1999. *J. Mass Spectrom.* 34: 105.

Gross, J. and Hillenkamp, F., 2000, In *Encyclopedia of Analytical Chemistry*, (Meyers, R.A., ed.) John Wiley and Sons, Inc., (2001) in press.

Hillenkamp, F., Karas, M., Beavis, R.C., Chait, B.T. 1991, *Anal.Chem.* **63**: 1193A.

Horneffer, V., Dreisewerd, K., Lüdemann, H.-C., Hillenkamp, F., Lage, M., Strupat, K., 1999, *Int. J. Mass Spectrom.* **185/186/187**: 859.

Iribarne, J.V.; Thomson, B.A., 1976, *J. Chem. Phys.* **64**: 2287.

Jensen, O.N., Podtelejnikov, A.; Mann, M., 1996, *Rapid Commun. Mass Spectrom.* **10**: 1371.

Karas, M. and Hillenkamp, F., 1988, *Anal.Chem.* **60**: 2299.

Karas, M., Bachmann, D., Hillenkamp, F., 1985, *Anal.Chem.* **57**: 2935.

Karas, M., Bahr, U., Ingendoh, A., Hillenkamp, F., 1989, *Angew.Chem.Int.Ed.Engl.* **28**: 760.

Karas, M., Ehring, H., Nordhoff, E., Stahl, B., Strupat, K., Hillenkamp, F., Grehl, M., and Krebs, B., 1993, *Org.Mass Spectrom.* **28**: 1476.

Karas, M., Glückmann, M., Schäfer, J., 2000, *J.Mass Spectrom.* **35**: 1.

Kebarle, P.; Tang, L., 1993, *Anal. Chem.* **65**: 972A.

Kussmann, M., Nordhoff, E., Rahbek-Nielsen, H., Haebel, S., Rossel-Larsen, M., Jakobsen, L., Gobom, J., Mirgorodskaya, E., Kroll-Kristensen, A., Palm, L., and Roepstorff, P., 1997, *J.Mass Spectrom.* **32**: 593.

Loo, J.A., 1997, *Mass Spectrom. Rev.* **16**: 1.

Mann, M.; Meng, C.K.; Fenn, J.B., 1989, *Anal. Chem.* **61**: 1702.

Nielen, M.W.F., 1999, *Mass Spectrom. Rev.* **18**: 309.

Paul, W., 1990, *Angew. Chem. Int. Ed. Engl.;* **29**, 739, also references therein.

Price, P., 1991, *J. Am. Soc. Mass Spectrom.* **2**: 336.

Shevchenko, A., Chernushevich, I., Ens, W., Standing, K.G., Thomson, B., Wilm, M., Mann, M., 1997, *Rapid Commun. Mass Spectrom.* **11**: 1015.

Shevchenko, A., Jensen, O.N.; Podtelejnikov, A., Sagliocco, F., Wilm, M.; Vorm, O.; Mortensen, P.; Shevchenko, A.; Boucherie, H.; Mann, M., 1996, *Proc. Natl. Acad. Sci. U.S.A.* **93**: 14440.
Shevchenko, A.; Wilm, M.; Vorm, O.; Mann, M., 1996, *Anal. Chem.* 68: 850.
Smith R.D.; Loo, J.A.; Edmonds, C.D.; Barinaga, C.J.; Udseth, H.R.,1990, *Anal. Chem.* **62**: 882.
Smith, R.D.; Bruce, J.E.; Qinyuan Wu; Lei, Q.P., 1997, *Chem. Soc. Rev.* **26**: 191.
Thomson, B.A.; Iribarne, J.V., 1979, *J. Chem. Phys.* **71**: 4451.
Vorm, O. and Mann, M., 1994, *J.Am.Soc.Mass Spectrom.* **5(11)**: 955,
Vorm, O.; Roepstorff, P.; Mann, M., 1994, *Anal.Chem.* **66**: 3281.
Wilm, M. and Mann, M., 1994, *Int. J. Mass Spectrom. Ion Proc.* **136**: 167.
Wilm, M. and Mann, M., 1996, *Anal. Chem.* **68**: 1.
Yates, J.R.,III, 1998, *J. Mass Spectrom.* **33**: 1.
Zaia, J.; Annan, R. S.; Biemann, K., 1992, *Rapid Commun. Mass Spectrom.* **6**: 32.
Zenobi, R. and Knochenmuss, R., 1999, *Mass Spectrom.Rev.* **17**: 337.

Adsorption Kinetics of Low Density Lipoprotein onto a Hydrophobic-Hydrophilic Gradient Surface

VLADIMIR HLADY
Department of Bioengineering, University of Utah, Salt Lake City, UT 84112, USA

1. INTERACTIONS BETWEEN PROTEINS AND SURFACES

A subtle interplay between polar and non-polar interactions governs protein stability in aqueous solutions (Branden and Tooze, 1991). It also plays a decisive role in determining the interfacial property of proteins. For example, many small globular proteins such as α-lactalbumin, lysozyme and others, can not pack all their hydrophobic residues inside the protein core: these residues are often found exposed to the aqueous solution (Haynes and Norde, 1995). This exposure is energetically costly as the water molecules in the vicinity of protein hydrophobic residues lose some of their entropy relative to the bulk water. The free energy of the system will decrease when the exposed hydrophobic residue binds to another hydrophobic entity, or when it makes contact with a hydrophobic surface.

The exposed hydrophobic residues in small globular proteins stand in contrast to the properties of large lipid-protein complexes known as lipoproteins. The lipoprotein particles are designed to transport water-insoluble lipids in the circulation. They pack very effectively hydrophobic components within the non-polar core, leaving thus the entire outside surface filled with polar lipid head-groups. In addition to polar head-groups, the outside lipoprotein surface contains one or more protein chains that wrap around the particle and further ensure particle stability. These proteins have another important function: they serve as a targeting moiety that will be recognized by respective cell membrane receptors (Gotto, 1987).

There are several different types of lipoproteins present in circulation. Two of those, high (HDL) and low density lipoproteins (LDL) are present in

Supramolecular Structure and Function 7, edited by Pifat-Mrzljak.
Kluwer Academic/Plenum Publishers, New York, 2001.

circulation at concentration that at times exceeds 1 g/L. Together with other ten proteins whose concentration is larger than 1 g/L, these proteins, by virtue of their high concentration, tend to dominate early process of blood contact with surfaces. Interest in lipoproteins primarily arises from their association with coronary artery disease. The popular interpretation of LDL as "bad" and HDL as "good" lipoprotein is based on their use as risk markers for atherogenesis and coronary artery disease. An interrelation between an atherosclerotic plaque formation and arterial thrombosis has been shown to be due to the action of lipoprotein (a), Lp(a), a variant of LDL that in its structure contains an additional protein, apolipoprotein(a), homologous to plasminogen (Utermann, 1989, Miles *et al.*, 1989, Hajjar *et al.*, 1989). It is interesting that both adsorbed LDL and Lp(a) have been shown to strongly promote a procoagulant state (Simon *et al.*, 1991).

Blood protein adsorption man-made biomaterials is still as a major problem in hemocompatibility of biomaterials (Andrade and Hlady, 1986). The adsorbed protein layer is likely to induce activation of coagulation cascade and/or platelet adhesion and thrombus formation. Various surface modifications aimed at reducing or eliminating protein adsorption are the major route in improving hemocompatibility of biomaterials (Andrade and Hlady, 1986). Hydrophilic surfaces are known to typically adsorb less protein (Andrade, 1985). Unlike hydrophilic surfaces, biomaterials with hydrophobic surfaces have been shown to strongly adsorb albumin and other proteins. The so-called *gradient surfaces*, originally described by Elwing *et al.* (1987), (Hlady, 1991, Golander *et al.*, 1990, Lin and Hlady, 1994) proved to be useful tools in evaluating the effect of surface hydrophilic/hydrophobic balance on protein adsorption. We have used the surface gradient of hydrophobic octadecyldimethylsilyl (C18) groups to study adsorption of human serum albumin (HSA), protein with the highest plasma concentration. HSA adsorption was found to increase with increasing surface density of C18 groups (Lin and Hlady, 1994). The HSA surface affinity levels at approximately 16 - 25 C18 surface groups per albumin molecule; a surface with the C18 surface density higher than that is "seen" by HSA molecules as being generically hydrophobic.

Several research groups, including ours, have studied the interfacial behavior of human lipoproteins (Breemhaar *et al.*, 1984, Shen, 1985, Dong *et al.*, 1987, Hlady *et al.*, 1989, Poot *et al.*, 1990, Healy and Ducheyne, 1991, Ho and Hlady, 1997, Ho *et al.*, 1997, Hlady and Ho, 2001). We have utilized the gradients of C18 groups on hydrophobic silica (i.e. C18-silica gradient surface) to investigate the effect of surface hydrophobicity on the adsorption of LDL from a single protein solution and from different binary protein solution mixtures (i.e. LDL + HSA, LDL + IgG) using the spatially-resolved total internal reflection fluorescence (TIRF) and scanning force microscopy

(SFM) techniques(Ho and Hlady, 1997, Ho *et al.*, 1997, Hlady and Ho, 2001). By performing two identical TIRF adsorption experiments, one with protein-labeled LDL, the other with lipid-labeled LDL, we also confirmed that both protein and lipid components of LDL are together adsorbed on the C18-silica gradient surface (Ho and Hlady, 1997).

In this contribution we show how the analysis of the experimental LDL adsorption kinetics can be used to find the lipoprotein affinity for adsorption onto surfaces with varying hydrophobicity. A more detailed description of the protein adsorption kinetics analysis can be found elsewhere (Hlady *et al.*, 2000).

2. QUANTITATIVE ANALYSIS OF PROTEIN ADSORPTION

The following four processes are thought to contribute to experimentally observed protein adsorption kinetics:

i) transport towards the interface,
ii) attachment at the interface,
iii) eventual structural rearrangements in the adsorbed state,
iv) detachment from the interface, and
v) transport away from the interface.

In principle, each of these steps can control the overall rate of the adsorption process.

2.1 Transport towards the interface

We first consider the mass-transport of protein molecules in a solution that flows through a rectangular flow channel (Lok *et al.*, 1983, Lok *et al.*, 1983, Kim *et al.*, 1993). The analysis starts from the analytical solution to the equivalent heat-transfer problem that has been solved by Leveque (1928). It is assumed that the walls of the flow channel act as sinks for all arriving molecules so that the the flux of protein molecules to the surface, J_p, (kg m^{-2} s^{-1}), is:

$$J_p = [\Gamma(4/3)]^{-1}\, 9^{-1/3}\, (6\, q\, b^{-2}\, w^{-1}\, l^{-1}\, D_p^{-1})^{1/3}\, D_p\, c_p \qquad (1)$$

where D_p (m^2 s^{-1}) is the diffusion coefficient of the protein, c_p (kg m^{-3}) is protein solution mass concentration, $[\Gamma(4/3)]$ is the gamma function of 4/3, l (m) is the distance along the rectangular flow channel measured from its entrance, b (m) is the thickness of the flow channel, q (m^3 s^{-1}) is the volume flow rate of protein solution, and w (m) is the width of the flow channel. In the case of pure transport-limited adsorption and no desorption, J_p is equal to the rate of protein adsorption to the surface, $d\Gamma_p(t)/dt$ (kg m^{-2} s^{-1}).

In the case of C18-silica gradient surfaces (to be described below), it is possible to design the experiment in such a way that the protein adsorption to one region of the gradient surface is initially rate-limited by the transport to the surface, while in the other regions of the gradient surface the rate-limiting process controlled by other mechanism. Because any protein adsorption kinetics cannot proceed faster than the rate of the supply of protein molecules to the adsorbing surface allows (i.e. *transport-limited* adsorption), we have often utilized the mass-transport adsorption regime to calibrate the TIRF-measured fluorescence intensity in the terms of protein surface concentration (Lin and Hlady, 1994, Lin and Hlady, 1995).

In the case of the transport-limited protein adsorption and very slow or no desorption, the Leveque equation (eq. 1) predicts that the protein surface concentration increases linearly with time. The rate of transport-limited adsorption is proportional to $D_p^{2/3}$, c_p, and the wall shear rate, $\gamma^{1/3}$, which is defined as:

$$\gamma = 6\, q\, b^{-2} w^{-1} \tag{2}$$

2.2 Modeling protein attachment to and detachment from a surface

A simple protein adsorption model, to which we will fit the experimental LDL adsorption data in the subsequent sections, comprised only two processes: attachment at the interface, and detachment from the interface (i.e. adsorption and desorption). The model also computes the concentration of protein at an interface. This is done by calculating the rate of diffusion through a concentration boundary layer (often called the "unstirred" layer) of a constant thickness. The order of magnitude thickness of this layer, δ, is:

$$\delta = O\,(D_p\, l\, \gamma^{-1})^{1/3} \tag{3}$$

It has been shown that a linear protein concentration profile across the "unstirred" layer develops rather rapidly (Corsel *et al.*, 1986). The model balances the rates of protein attachment and detachment to find the overall adsorption rate, $d\Gamma_p(t)/dt$:

$$d\Gamma_p(t)/dt = k_{on}\,[\Gamma_p^{max} - \Gamma_p(t)]\; c_p(0,t) - k_{off}\Gamma_p(t) \tag{4}$$

where k_{on} (in units of concentration^{-1} time^{-1}) and k_{off} (in units of time^{-1}) are the intrinsic adsorption and desorption rate constants, respectively, Γ_p^{max} is the maximum protein surface mass concentration, $\Gamma_p(t)$ is the protein surface mass concentration at time t, and $c_p(0,t)$ is a function that describes the protein concentration at distance $y = 0$ from the adsorbent surface at time t, where y is measured normal to the surface. Equation 4 can be solved with the following boundary conditions (Corsel *et al.*, 1986):

(a) the rate of adsorption is linearly dependent to the protein concentration gradient right next to the surface, $\partial c_p(0,t)/\partial y$:

$$d\Gamma_p(t)/dt = D_p\ [\partial c_p(0,t)/\partial y] \tag{5}$$

and

(b) protein concentration at distances δ or greater from the surface equal the protein solution concentration:

$$c_p(\delta,t) = c_p \tag{6}$$

The model assumes that a linear protein concentration gradient in the "unstirred" layer exists since
t = 0 (Corsel *et al.*, 1986). The boundary condition (a) (eq. 5) is then re-cast as:

$$d\Gamma_p(t)/dt = D_p\ [(c_p - c_p(0,t))\ \delta^{-1}] \tag{7}$$

or:

$$c_p(0,t) = c_p - (\delta D_p^{-1})\ d\Gamma_p(t)/dt \tag{8}$$

By substituting eq. 8 into eq. 4, one finds that:

$$d\Gamma_p(t)/dt = k_{on}^{app}\ [\Gamma_p^{max} - \Gamma_p(t)]\ c_p - k_{off}^{app}\Gamma_p(t) \tag{9}$$

where k_{on}^{app} and k_{off}^{app} are defined as the apparent adsorption rate constant and desorption rate constant:

$$k_{on}^{app} = k_{on}\ D_p\ (D_p + \delta \cdot k_{on}\ [\Gamma_p^{max} - \Gamma_p(t)])^{-1} \tag{10}$$

$$k_{off}^{app} = k_{off}\ D_p\ (D_p + \delta \cdot k_{on}\ [\Gamma_p^{max} - \Gamma_p(t)])^{-1} \tag{11}$$

The model can account for any adsorption cooperativity by allowing the intrinsic adsorption and desorption rate constants, k_{on} and k_{off}, to be defined as empirical exponential functions of protein surface concentration, Γ:

$$k_{on} = k_1 \, e^{-\alpha\Gamma} \tag{12}$$

$$k_{off} = k_{-1} \, e^{\beta\Gamma} \tag{13}$$

where k_1 is the initial intrinsic adsorption rate constant, k_{-1} is the initial intrinsic desorption rate constant, and α and β are the adsorption and desorption "cooperativity" constants (Corsel *et al.,* 1986).

A simple computer program can be used to numerically solve eq. 9 in order to simulate the adsorption kinetics of protein using a set of the parameters: Γ_p^{max}, D_p, δ, k_1, k_{-1}, α, β, and $c_p(20)$. When the experimental results are to be compared with the model predictions the computer program searches for the parameters k_1, α, and β that will give the best fit. This fitting process is often preceded by finding k_{-1} using experimentally measured desorption as described in the next section.

2.2.1 Finding the desorption rate constant

In a typical protein adsorption experiment, one measures the desorption of adsorbed protein by flowing buffer solution with no protein present over the adsorbed protein layer. Since ($c_p = 0$), eq. 9 can be written as:

$$d\Gamma_p(t)/dt = -\, k_{off}^{app}\Gamma_p(t) \tag{14}$$

At high protein surface coverage, (i.e. when $[\Gamma_p^{max} - \Gamma_p(t)] \rightarrow 0$), there are no free binding sites available for re-adsorption of desorbed molecules so that:

$$k_{off}^{app} \sim k_{off} \tag{15}$$

In this case, the desorption segment of the experiment can be modeled as:

$$d\Gamma_p(t)/dt = -\, k_{off}\Gamma_p(t) \tag{16}$$

or, in an integral form as:

$$\Gamma_p(t) = \Gamma_p(t_{d\text{-}beg}) \, e^{-t\, koff} \tag{17}$$

where $\Gamma_p(t_{d\text{-}beg})$ and $\Gamma_p(t)$ are the protein surface mass concentrations at the beginning of the desorption segment ($t_{d\text{-}beg}$) and at the time t, respectively.

2.2.2 Modeling of the adsorption process

Once the desorption rate has been found, one proceeds with the modeling of the adsorption segment of the experiment. First, the approximate thickness of the "unstirred" layer, δ, is calculated using eq. 3. The magnitude of δ depends on the actual flow cell dimensions, position along the flow chamber and flow rate used in the experiments. The other parameter needed is Γ_p^{max} . In principle, the fitting program can find its "best" Γ_p^{max} value. Alternatively, one can measure Γ_p^{max} directly. Potential problem with any measured Γ_p^{max} parameter is related to the degree of uncertainty about the conformational state of adsorbed protein molecules. Namely, if the adsorbed protein undergoes a conformational change, the measured Γ_p^{max} may represent the adsorption saturation of denatured protein, a molecular species whose adsorption kinetics were not measured in the experiment. The fitting of the experimental data is typically carried out with k_1 and α as the only two adjustable parameters (recall that the parameters k_{off}, or k_{-1} and β, should have been found from the desorption segment of the experiment).

3. ADSORPTION OF LDL TO HYDROPHILIC-HYDROPHOBIC GRADIENT SURFACE

3.1 Materials and methods

3.1.1 Isolation of human low density lipoprotein

Human low density lipoprotein (LDL) was isolated from pooled human blood using the method of ultracentrifugation flotation (Mills *et al.,* 1976). The purity of LDL fraction was checked by one dimensional polyacrylamide gel electrophoresis. Aliquots of LDL (1.5 g/L apolipoprotein B, or 6.0 g/L LDL) were frozen at -20°C, respectively. For each adsorption experiment, a desired amount of LDL was thawed. Concentration of the protein component of LDL was determined by a modified Lowry method (Lowry *et al.,* 1951). The total concentration of the LDL in any solution was estimated by assuming 25% of protein and 75% of lipid in each LDL particle (Soutar and Myant, 1987).

3.1.2 Labeling of proteins

Fluorescein isothiocyanate isomer I (FITC, Aldrich) was covalently bound to LDL following the method of Coons *et al.* (1942). The degree of labeling was determined from the absorbances measured at 280 and 496 nm (Lin and Hlady, 1995). The degree of labeling for FITC-LDL was found to be 1.5. ^{125}I-labeled LDL was prepared by using the modified ICl iodination protocol (Shepherd *et al.*, 1976). The reaction was performed at pH 10 to minimize lipid labeling. The free ^{125}I was removed from the labeled LDL using the PD-10 (Pharmacia) column.

3.1.3 Preparation of C18-silica gradient surface

The C18-silica gradient surface was prepared using the two phases diffusion method (Elwing *et al.*, 1987). A high density solvent used was a solution mixture of 125 mL dichloromethane (density, d = 1.325 g cm^{-3}, EM Science), 1 mL pyridine (J.T. Baker) and 2 mL octadecyldimethylchlorosilane (ODS, Aldrich). It was layered below a low density solvent, 150 mL p-xylene (d = 0.86 g cm^{-3}, Fluka), in a container with six clean silica plates (size 2.5 × 7.5 cm). The high density solvent and the low density solvent counter-diffuse in each other forming an ODS concentration gradient at the interface between the two phases. The silanization reaction was allowed to proceed for 4 hours. The C18-silica gradient surfaces were stored in a glass jar filled with the deionized water. The C18 gradient was characterized by measuring the water contact angle as a function of the gradient position.

3.1.4 LDL adsorption experiments

LDL adsorption experiments were performed using the dual channel TIRF technique described in full details elsewhere (Hlady, 1991). Two parallel experiments were performed using the same C18-silica gradient surface. The concentrations of lipoprotein was equal to the 1/100 of its respective concentrations in normal plasma (c_{LDL} = 4 × 10^{-2} kg m^{-3}). The protein solution in Dulbecco phosphate buffer (DPBS, 0.05 M phosphate, 0.147 NaCl, 0.90 mM $CaCl_2$, 0.88 mM $MgCl_2$, 2.70 mM KCl, pH 7.4) was flowed into the flow channel from the hydrophilic end with the rate of 0.84 mL/min initially displacing the buffer solution. The adsorption-desorption cycle included 11 minutes of adsorption and another 11 minutes of desorption time. The fluorescence intensity profile was recorded along the gradient each second using a CCD camera (Ho and Hlady, 1997).

The autoradiography technique was used for the quantitation of the TIRF experiments, i.e. for obtaining the information about the spatial distribution of the adsorbed protein amount along the gradient (Lin *et al.*, 1992).

Autoradiography only provided the final adsorbed amount of lipoprotein along the gradient at the end of the adsorption experiment. In order to quantify TIRF experiments, the autoradiography experiments were carried out with ^{125}I-labeled LDL using conditions that were otherwise identical to the conditions of TIRF experiments. After the full adsorption-desorption cycle, the adsorbed ^{125}I-labeled LDL on the gradient surface was fixed with 3 mL freshly prepared 0.6% glutaraldehyde solution in DPBS. A plate with a set of known amounts of ^{125}I-labeled LDL was prepared separately in order to serve as an autoradiography calibration standard for quantifying the adsorbed protein mass. The autoradiographs were obtained by placing the gradient silica plate with adsorbed ^{125}I-labeled proteins in contact with a 5×7 in. photosensitive film (X-OMAT AR, Kodak) in a light-tight cassette. The film was exposed at a low temperature (-70°C) for 21 days. No intensifying screens were used. After 21 days, the exposed film was processed in an automated developer system, the optical density was recorded by a custom-built densitometer and converted in the adsorbed protein mass/area as previously described (Lin *et al.*, 1992, Ho *et al.*, 1991).

3.2 Modeling LDL adsorption from experimental adsorption kinetics

3.2.1 Adsorption of FITC-LDL onto the hydrophilic and hydrophobic ends of the gradient surface

The autoradiography calibrated FITC-LDL adsorption and desorption kinetics onto the hydrophilic (silica) end of the gradient surface are shown in Figure 1a[1]. The flux of FITC-LDL (≈ transport-limited adsorption rate) was computed from eq. 1 using $q = 1.4 \times 10^{-8}$ m^3 s^{-1}, $b = 5 \times 10^{-4}$ m, $w = 5 \times 10^{-3}$ m, $l = 2.8 \times 10^{-2}$ m, $D_{\underline{LDL}} = 1.8 \times 10^{-11}$ m^2 s^{-1} (17), and $c_{\underline{LDL}} = 4 \times 10^2$ kg m^{-3} and is shown as a straight line in Figure 1a. The slope was found to be in good agreement with the TIRF adsorption results calibrated using

---

[1] Mass concentration SI units are used instead of the molar (or molecular) units. Accordingly, Γ_p and c_p will have the units of mass area^{-1} and mass volume^{-1}, respectively. The conversion of mass concentration units into molar (or molecular) units requires that the molecular mass of protein, M_p, is known, which may not always be the case.

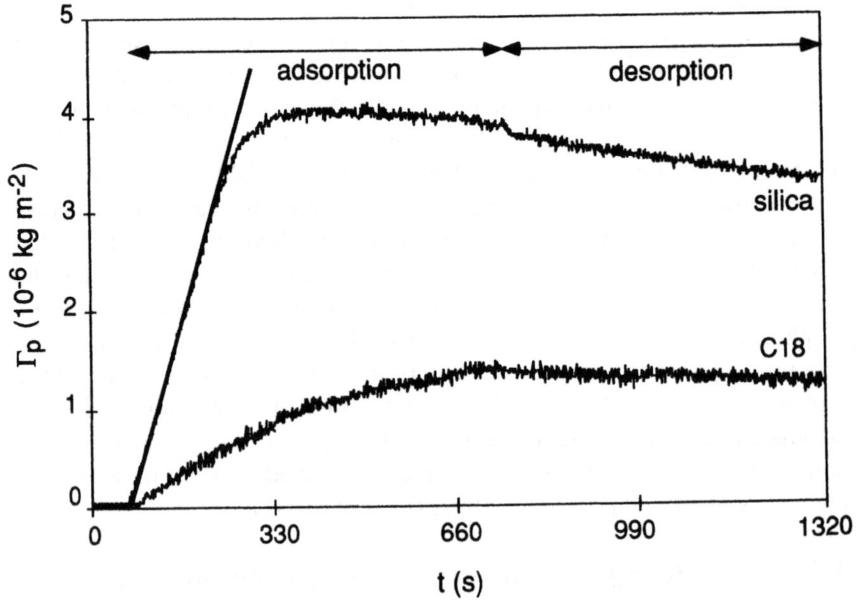

Figure 1a. Calibrated FITC-LDL adsorption and desorption kinetics onto the hydrophilic (silica) and hydrophobic (C18) region of the C18-silica gradient surface. The "transport-limited adsorption" slope, shown as a straight line, was computed from eq. 1.

autoradiography. The flux of FITC-LDL to the other regions of the C18-silica gradient surface scaled by a small factor of $l^{1/3}$[2]. Figure 1b shows the FITC-LDL adsorption kinetics onto the hydrophilic (water contact angle, $\theta_{adv} = 0°$) and hydrophobic (water contact angle, $\theta_{adv} = 104°$, fractional C18 surface coverage, $\Theta_{C18} = 0.9$ (Ho and Hlady, 1997)) ends of the gradient surface. Before the fitting of the experimental LDL adsorption, k_{-1} was found first by fitting the desorption segments to the eq. 17. The magnitude of δ depended on the actual flow cell dimensions and flow rate used in the experiments: δ amounted to 3.1×10^{-5} m at the hydrophilic region and 3.6×10^{-5} m at the hydrophobic region of the C18-silica gradient surface, respectively. The difference between the two values is due to the distance between these two regions in the TIRF flow cell. Γ_p^{max} has been set to 4×10^{-6} and 3×10^{-6} kg LDL m^{-2} for the hydrophilic and hydrophobic regions, respectively. These two Γ_p^{max} values agreed very well

[2] In the calibration of TIRF-measured adsorption kinetics it was assumed that the fluorescence quantum yield of adsorbed FITC-LDL remains constant during the process of adsorption. Furthermore, the quantum yield of fluorescence of adsorbed FITC-LDL was assumed to have identical magnitude along the C18-silica gradient surface.

with the previous study of LDL adsorption onto the hydrophobic and hydrophilic silica (Hlady *et al.*, 1989). The parameters used to achieve the fits shown in Figure 1b are listed in Table 1.

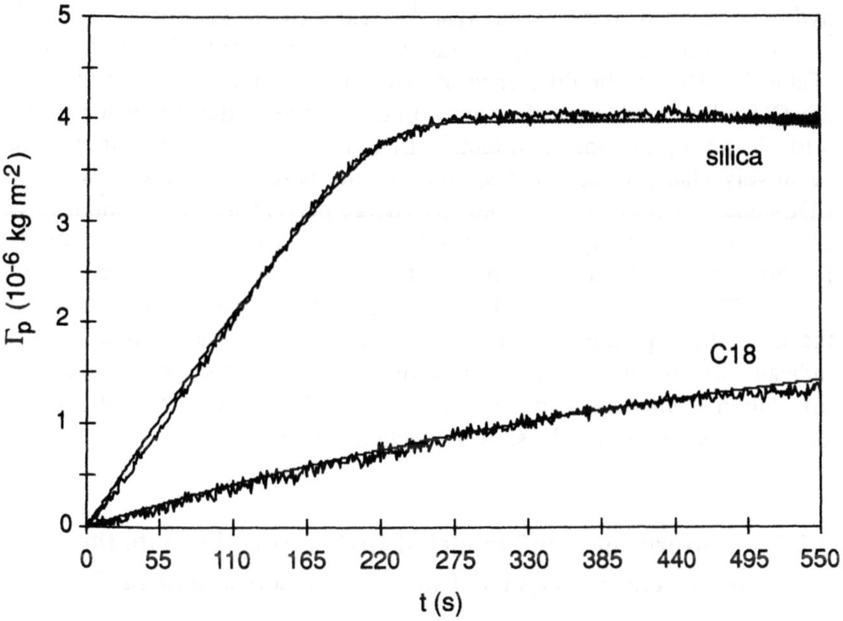

Figure 1b. Comparison between the model prediction (solid line) and the experimental results (calibrated TIRF trace) of FITC-LDL adsorption onto the hydrophilic and hydrophobic region of the C18-silica gradient surface.

Table 1. Adsorption parameters obtained by fitting the experimental LDL adsorption (Figure 1b) to the model given by eqs. 9 - 13.

	hydrophilic silica	C18-covered silica
$\theta_{adv}/^o$	0	104
Θ_{C18}	0	0.9
δ/m	3.1×10^{-3}	3.6×10^{-3}
α/kg^{-1}m^2	-2.0×10^5	2.5×10^5
β/kg^{-1}m^2	0	0
Γ_{LDL}^{max} / kg m^{-2}	4.0×10^{-6}	3.0×10^{-6}
k_1/kg^{-1}m^{3}s^{-1}	7.5×10^{-1}	4.5×10^{-2}
k_{-1}/s^{-1}	2.9×10^4	2.3×10^4
(K/M^{-1})*	2.6×10^3	2.0×10^2
K/kg^{-1}m^3	(6.5×10^9)	(4.9×10^8)

*The *K* values in units of M^{-1} were calculated by assuming $M_{LDL} = 4.35 \times 10^{-21}$ kg (Hlady *et al.*, 1989).

The apparent LDL adsorption equilibrium constant, K_{LDL} (Table 1), was found from the k_1/k_{-1} ratio.[3] FITC-LDL adsorption affinity for the negatively charged, hydrophilic silica surface, K_{LDL}^{silica}=2.6×10^3 m^3 kg^{-1} (or, 6.5×10^9 M^{-1}), was by an order of magnitude larger than the affinity for the hydrophobic C18-silica surface, K_{LDL}^{C18} = 2.0 × 10^2 m^3 kg^{-1} (or, 4.9 × 10^8 M^{-1}). The two respective desorption rate constants were only slightly different (Table 1). Hence, the difference in K_{LDL}s between the different regions of the C18-silica gradient surface was primarily due to the difference in the intrinsic adsorption rate constants. LDL is known to bind strongly to negatively charged adsorbents; some of the best commercially available LDL-apheresis devices are negatively charge polyelectrolytes (Ikonomov *et al.*, 1992). Accordingly, the electrostatic interactions seem to be responsible for the higher LDL adsorption affinity for the negatively charged silica surface. The mechanism of LDL adsorption onto the hydrophobic region of the C18-silica gradient surface is less obvious. The adsorption kinetics indicates an energy barrier to adsorption. It is possible that an improper orientation of the LDL particle during some collisions with the hydrophobic surface is the reason for the slower attachment rate.

3.2.2 "Adsorption sum" model of LDL adsorption onto the intermediary regions of the C18-silica gradient surface

The LDL adsorption kinetics indicated that the adsorption onto the hydrophilic silica surface takes place via different mechanism than the adsorption onto the hydrophobic C18 silica. To answer the question whether the LDL adsorption on a mixed hydrophobic-hydrophilic surface behaves as a simple addition of two adsorption processes, or the adsorption onto neighboring hydrophobic and hydrophilic sites affect each other, the experimental LDL adsorption to the silica-C18 gradient region was first fitted to an "adsorption sum" model(20). The model adds the adsorption on the hydrophobic (subscript $_{-C18}$) and hydrophilic (subscript $_{-sil}$) surface sites, respectively, and weights each adsorption process according to the probability of finding particular adsorption site:

[3]In order to estimate K_{LDL} in M^{-1} units, one has to know the molecular mass of LDL molecule, M_{LDL}, and convert k_1 *(mass concentration)*$^{-1}$ *time*$^{-1}$ units into *(molar concentration)*$^{-1}$ *time*$^{-1}$ units. While the exact molecular mass of LDL particle is often not known, the conversion has been performed here by assuming M_{LDL} to be 4.35 × 10^{-21} kg (Hlady *et al.*, 1989).

$$d\Gamma_p(t)/dt = (1 - \Theta_{C18})\, k_{on\text{-}sil}{}^{app}\, [\Gamma_{p\text{-}sil}{}^{max} - \Gamma_{p\text{-}sil}(t)]\, c_p - k_{off\text{-}sil}{}^{app}\Gamma_{p\text{-}sil}(t) + \Gamma_{C18}\, k_{on\text{-}C18}{}^{app}\, [\Gamma_{p\text{-}C18}{}^{max} - \Gamma_{p\text{-}C18}(t)]\, c_p - k_{off\text{-}C18}{}^{app}\Gamma_{p\text{-}C18}(t) \quad (18)$$

The comparison between the experimental adsorption kinetics and the predictions from the "adsorption sum" model is shown in Figure 2. Included in the same figure is the hypothetical LDL adsorption fit for a fully covered C18 surface.[4] The inspection of Fig. 2 shows that the "adsorption sum"

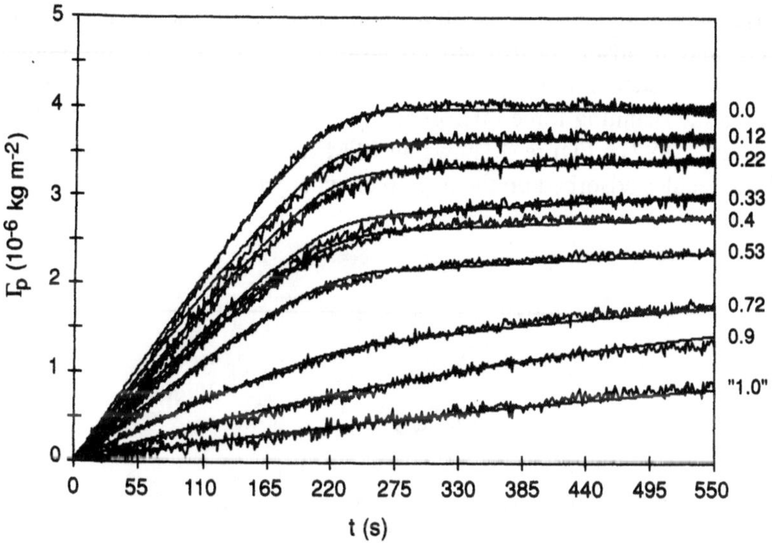

Figure 2. Comparison between the modeled adsorption (solid lines) and the experimental results (calibrated TIRF trace) of FITC-LDL adsorption onto the intermediary regions of the C18-silica gradient surface. The surface coverage of C18 groups, Θ_{C18}, is indicated. The adsorption of LDL onto a hypothetical, fully covered C18 surface (i.e. $\Theta_{C18} = 1.0$) was obtained by linear extrapolation of results recorded for $\Theta_{C18} = 0.9$.

model (eq. 18) fits very well with the experimental LDL adsorption for the C18 surface coverage close to 1 or to 0. The fit worsens as one moves towards the middle of the gradient where $\Theta_{C18} \rightarrow 0.5$, especially in reaching the adsorption steady-state level. In spite of that, one concludes that this simple additive two sites model (eq. 18) that accounted for the fractional surface coverage C18 groups in the gradient region, is capable of predicting adsorption data in the intermediary gradient regions quite well. It also appears that the adsorption processes on the two types of surface sites (i.e.

[4] The adsorption parameters for LDL adsorption onto a fully covered C18 (i.e. $\Theta_{C18} = 1.0$) surface were obtained by linear extrapolation.

silica vs. C18) are not affecting each other. This finding may indicate that the size of C18 islands on the silica surface is larger than the size of the LDL molecule.

3.2.3 Cooperativity of LDL adsorption onto negatively charged surfaces

A small positive adsorption cooperativity ($\alpha = -2 \times 10^5$ m^2 kg^{-1}, Table 1) was introduced in order to fit the experimental FITC-LDL adsorption onto the hydrophilic silica.[5] The positive cooperativity implies that the adsorbed proteins like to adsorb nearby already adsorbed molecules. A corroborating piece of evidence for the positive cooperativity in LDL adsorption was found in the scanning force microscopy (SFM) LDL adsorption experiments (Ho *et al.*, 1997). Figure 3 shows an SFM image (5 × 5 μm^2 area) of the LDL molecules adsorbed on the negatively charged mica surface.

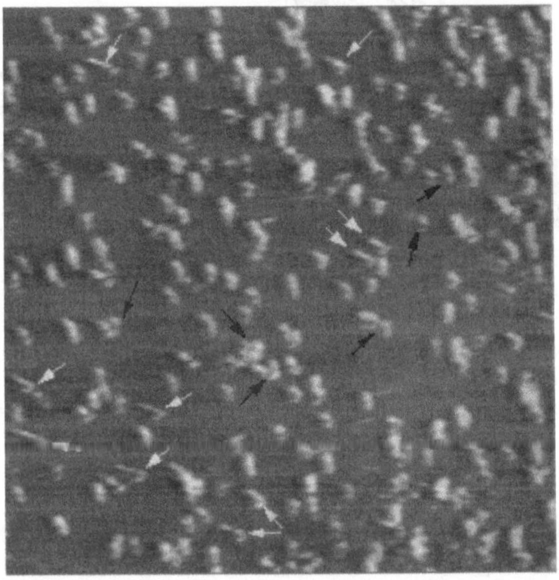

Figure 3. SFM image (5 × 5 μm^2 area) of LDL adsorbed on freshly-cleaved mica. Adsorption was allowed to take place for 15 minutes, mica was then rinsed and imaged by SFM in air.

Quite a few of adsorbed LDL molecules are found aggregated in tetrameric "L"- (indicated by white arrows) and "clover"-shaped (indicated by black

[5] Note that the sign of the adsorption cooperativity constant, α, is opposite to the cooperativity itself.

arrows) patterns as well as in clusters of dimers and trimers. The *in situ* SFM imaging of LDL adsorption on mica indicated that the aggregation occurred during the adsorption processes and was not present in solution (Ho *et al.*, 1997).

The best fit for the LDL adsorption on hydrophobic surface required that α is set to $2.5 \bullet 10^5$ m^2 kg^{-1} (Table 1). One interpretation of this negative adsorption cooperativity is based on a repulsive potential that exists around each adsorbed protein molecule. Such a potential makes subsequent attachments of molecules in the neighborhood of the adsorbed molecule more difficult. It is possible that positive charges on LDL apo B-100 protein are the source of this negative cooperativity. In the case of negatively charged hydrophilic silica (and mica) the surface charge can effectively screen the positive B-100 charge and, by doing so, allow LDL molecule to aggregate (Figure 3). One can speculate that the aggregation of LDL at negatively charged surfaces could be related to pathological formation of lipoprotein deposits in the cardiovascular system. The molecular origin of this positive cooperativity, however, remains to be unknown and needs to be further studied and elucidated.

4. SUMMARY

Experimental adsorption kinetics of human LDL onto a C18-silica gradient surface were interpreted with the help of a simple adsorption/desorption model that accounted for the changes in lipoprotein concentration at the adsorbing surface during the adsorption experiment. In actual experiments, the FITC-LDL flow rate and concentration were adjusted in such a way that the adsorption rate was transport-limited on the negatively charged silica. This experimental feature confirmed the conversion of the fluorescence from adsorbed LDL into the adsorbed LDL mass per area. A simple adsorption model comprising only the adsorption and desorption processes and allowing for the effects of cooperativity was used to calculate the adsorption and desorption rate constants, and the adsorption cooperativity for each end of the C18-silica gradient surface. Interestingly, the LDL adsorption onto the intermediary regions of the C18-silica gradient surface could be accurately predicted by a simple "adsorption sum" model in which the adsorption onto each type of surface site was weighted according to the probability of finding particular adsorption site. The additive nature of LDL adsorption onto the hydrophilic (silica) and hydrophobic (C18) surface sites may be an indication that the C18 sites were in the form of islands larger than the LDL molecule.

ACKNOWLEDGMENT

All blood donors are thanked for providing plasma from which low density lipoprotein was isolated. The work on LDL adsorption onto C18-silica gradient surfaces comes from the Ph.D. thesis of Dr. C.-H. Ho (1992). The SFM data were obtained by Drs. D. W. Britt and C.-H. Ho (1997). This work was financially supported by NIH grant (R01 NIH-44538) and the Center for Biopolymers at Interfaces, University of Utah.

REFERENCES

Andrade, J. D., 1985, *in Surface and Interfacial Aspects of Biomedical Polymers. 2. Protein Adsorption*, (Andrade, J. D., Ed.), Plenum Press, New York, pp. 1 - 79.

Andrade, J. D. and Hlady, V., 1986, *Adv. Polym. Sci.* **79**: 1.

Branden, C. and Tooze, J., 1991, *Introduction to protein structure*, Garland Publishing, New York.

Breemhaar, W., Brinkman, E., Ellens, D. J., Beugeling, T. and Bantjes, A., 1984, *Biomaterials* **5**: 269.

Coons, A. M., Crech, H. J., Jones, R. N. and Berliner, E. J., 1942, *J. Immunology* **45**: 59.

Corsel, J. W., Willems, G. M., Kop, J. M. M., Cuypers, P. A. and Hermens, W. T., 1986, *J. Colloid Interface Sci.* **111**: 544.

Dong, D. E., Andrade, J. D. and Coleman, D. L., 1987, *J. Biomed. Mater. Res.* **21**: 683.

Elwing, H., Welin, S., Askendahl, A., Nilsson, U. and Lundstrom, I., 1987 *J. Colloid Interface Sci.* **119**: 203.

Golander, C.-G., Lin, Y.-L., Hlady, V. and Andrade, J. D., 1990, *Colloids and Surfaces* **49**: 289.

Gotto, A. M. Jr., (Ed.), 1987, *Plasma Lipoproteins*, Elsevier, Amsterdam, pp. 405.

Hajjar, K. A., Gavish, D., Breslow, J. L. and Nachman, R. L., 1989, *Nature* **339**: 303.

Haynes, C. A. and Norde, W., 1995, *J. Colloid Interface Sci.* **169**: 313.

Healy, K. E. and Ducheyne, P., 1991, *ASAIO Trans.* **37**: M150.

Hlady, V., 1991, *Appl. Spectroscopy* **45**: 246.

Hlady, V. and Ho, C.-H., 2001, *Mat.-Wiss. u. Werkstofftech.* in press.

Hlady, V., Ho, C.-H. and Britt, D. W., 2000, in *Interfacial Dynamics*, (N. Kallay, Ed), Surfactant Science Series, Vol. **88**, M. Dekker Inc., New York, 2000, pp. 405 - 418.

Hlady, V., Rickel, J. and Andrade, J. D., 1989, *Colloids and Surfaces* **34**: 171.

Ho, C.-H., 1992, Ph.D. Dissertation, University of Utah, Salt Lake City.

Ho, C.-H. and Hlady, V., 1997, in *Protein at Interfaces 1994*, (Brash, J. and Horbett, T., Eds.), ACS, Washington, D.C., pp. 371-384.

Ho, C.-H., Britt, D. W. and Hlady, V., 1997, *J. Molecular Recognition* **9**: 444.

Ho, C.-H., Hlady, V., Nyquist, G., Andrade, J. D. and Caldwell, K. D., 1991, *J. Biomed. Mater. Res.* **25**: 423.

Ikonomov, V., Samtleben, W., Schmidt, B., Blumenstein, M. and Gurland, H., 1992, *J. Int. J. Artif. Organs* **15**: 312.

Kim, D., Chao, W. and Bessinger, R. L. *J. Colloid Interface Sci.*, 1993, **159**: 1.

Leveque, M., 1928, *Ann. Miner.* **13**: 284.

Lin, Y.-S. and Hlady, V., 1995, *Colloids Surf. B. Biointerfaces* **4**: 65.

Lin, Y.-S. and Hlady, V., 1994, *Colloids and Surfaces B: Biointerfaces* **2**: 481.

Lin, Y.-S., Hlady, V., and Janatova, J., 1992, *Biomaterials* **13**: 449.
Lok, B. K., Cheng, Y.-L. and Robertson, C. R., 1983, *J. Colloid Interface Sci.* **95**: 104.
Lok, B. K., Cheng, Y.-L. and Robertson, C. R., 1983, *J. Colloid Interface Sci.* **95**: 87.
Lowry, O. H., Rosebrough, N. J., Farr, A. L. and Randall, R. J., 1951, *J. Biol. Chem.* **193**: 265.
Miles, L. A., Fless, G. M., Levin, E. G., Scanu, A. M. and Plow, E. F., 1989, *Nature* **339**: 301.
Mills, G. L., Cane, P. A. and Weech, P. K., 1976, in *A guidebook to lipoprotein technique, Laboratory techniques in biochemistry and molecular biology*, (Burkon, R. H., and van Knippenberg, P. H., Eds.), Elsevier, Amsterdam, pp. 18.
Poot, A., Beugeling, T., van-Aken, W. G. and Bantjes, A., 1990, *J. Biomed. Mater. Res.* **24**: 1021.
Shen, B. W., 1985, *J. Biol. Chem.* **260**: 1032.
Shepherd, J., Bedford, D. K. and Mogan, H.G., 1976, *Clin. Chim. Acta* **66:** 453.
Simon, D. I., Fless, G. M., Scanu, A. M. and Loscalzo J., 1991, *Biochemistry* **30**: 6671.
Soutar, A. K. and Myant, N. B., 1987, *in International Review of Biochemistry, Chemistry of Macromolecules IIB*, (Offed, R. E., Ed.), University Park Press, Baltimore, pp. 55.
Utermann, G., 1989, *Science* **246**: 904.

Fluorescence Spectroscopic Studies on Structure and Function of Lipolytic Enzymes

ALBIN HERMETTER, BIRGIT MAYER, HUBERT SCHOLZE, ELFRIEDE ZENZMAIER, AND MARION GRAUPNER
SFB-Biocatalysis, Department of Biochemistry and Food Chemistry, Technical University Graz, Petersgasse 12/2, Graz, Austria

1. INTRODUCTION

Lipases and esterases are lipolytic enzymes hydrolyzing hydrophobic long- and short-chain carboxylic acid esters, respectively (Verger, 1997). Lipases play an important role in intracellular lipid metabolism and extracellular lipid degradation. Information on the biological function of esterases is rather scarce. In organic chemistry these proteins represent about 25 % of the biocatalysts currently used for stereo- and regioselective hydrolysis or esterification of synthetic substrates (Faber, 1992). Either crude enzyme preparations (e.g. pancreatic lipase) or a growing number of „pure" recombinant proteins are being employed for such purposes (Schmid and Verger, 1998). Knowledge of qualitative and quantitative enzyme composition is essential (Schmid and Verger, 1998) though not always available for crude enzymes. On the other hand, the fraction of active enzyme is an important parameter for (electrophoretically) pure (recombinant) lipases or esterases in monomeric or crystallized form. Both analytical requirements are not easy to meet using state of the art procedures (Schmid and Verger, 1998). Basic information on structure-function relationships of lipases and esterases has been obtained by X-ray crystallography (Anthonsen *et al.*, 1995). However, information on lipases in solution and their interactions with their (supramolecular) substrates is scarce. High-resolution NMR spectroscopy of these systems requires high protein concentrations and, at the same time, a minimum of protein self-association (Anthonsen *et al.*, 1994). Only the three-dimensional structure of cutinase has been solved by NMR to date. Fluorescence spectroscopy (Lacowicz, 1986) provides useful tools to study many analytical and structural problems associated with lipolytic enzymes in (micellar) solutions. Our recent developments in this field will be the subject of the following research report.

2. ACTIVITY AND STEREOSELECTIVITY OF LIPOLYTIC ENZYMES

2.1 Substrates

Fluorogenic 1(3)-*O*-alkyl-2,3(1,2)-diacylglycerols (Duque *et al.*, 1996) have been employed as substrates for the determination of activity and stereoselectivity of lipases (Figure 1, compounds B). The corresponding short-chain glycerolipids were used as fluorogenic substrates for esterases (Hermetter *et al.*, 1998) The respective compounds contain a fluorescent fatty acid and a quencher fatty acid (in glycerolipids) or a fluorescent alcohol component and a quencher fatty acid (in single-chain lipids). Their fluorescence is very low due to intramolecular resonance energy transfer

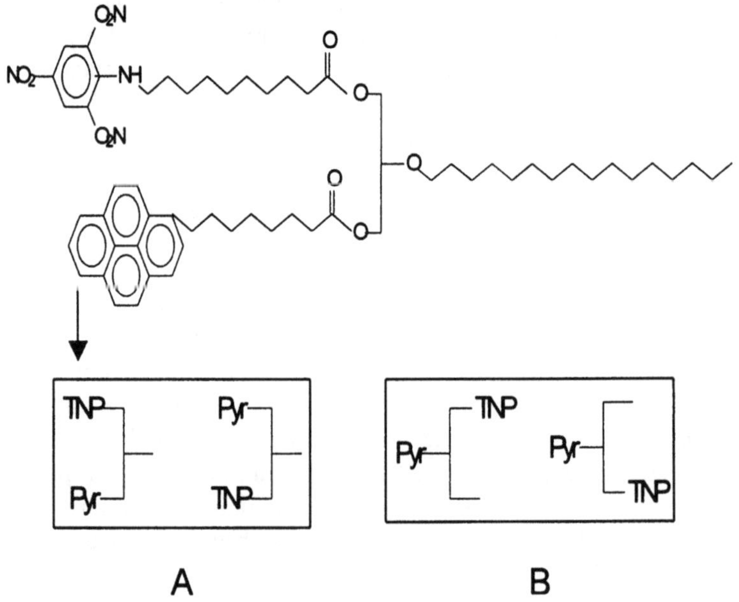

Figure 1. Fluorogenic substrates for lipolytic enzymes. A: 1(3)-Trinitrophenyl-amino-dodecanoyl-2-*O*-hexadecyl-3(1)-pyrenedecanoyl-*sn*-glycerol; B: 1(3)-Trinitrophenylamino-dodecanoyl-2-pyrenedecanoyl-3(1)-*O*-hexadecyl-*sn*-glycerol.

between the fluorophore (pyrene) and the quencher (trinitrophenylamine). Upon hydrolytic cleavage, e.g. by a lipase, the labeled hydrophobic chains are separated and fluorescence intensity increases. Thus, lipase activity is obtained from the time-dependent increase in fluorescence intensity. In addition, a new type of 1,3-diacyl-2-*0*-alkylglycerols (Figure 1, compounds A), based on the same fluorophore-quencher combination, has been developed. The respective compounds are hydrolyzed by lipases by one order of magnitude faster as compared to their 1,2(2,3) diacyl isomers and thus considerably improve sensitivity of the fluorescence lipase assay.

2.2 Substrate presentation

The described lipase substrates can be used in virtually all possible reaction media. For routine determination of lipase activity, a new, user-friendly form of the substrates was established (Hermetter, 1998). We found that the fluorogenic lipids can be solubilized by albumin in water, leading to defined lipid-protein-complexes with a size distribution identical to the lipid-free protein (dynamic light scattering measurements; M. Duque, O. Glatter and A. Hermetter, unpublished data). Ultrafiltration experiments showed that this complex was the only lipase substrate in the lipo-protein preparation. Larger lipid particles as potential lipase substrates could be excluded. The albumin-lipid complex in solution was lyophilized leading to a white powder which after dissolving in aqueous buffer represented a functional lipase substrate. Screening of appropriate additives led to the development of a stable lyophilisate which can be used for fast lipase assays that are unsurpassed in reproducibility and sensitivity.

2.3 Analytical applications of substrates

The new substrate(form)s are currently the basis for selective assays of lipoprotein lipase and hepatic lipase in post-heparin serum, newly developed by us for clinical diagnosis. The same lipids can also be used for the specific and sensitive detection of pancreatic lipase if a different buffer system is used, e.g. for screening of pancreatic lipase as a marker of pancreatitis. On the other hand, (crude) pancreatic lipase is one of the most frequently used enzymes for biocatalytic transformations of synthetic substrates in organic solvents (Faber, 1992). Activity of this enzyme can now selectively be determined in crude enzyme preparations using the new assay. Meanwhile assay conditions have been optimized for fluorescence plate readers and thus allow high throughput analyses of animal and microbial lipases in the routine laboratory. Recombinant esterases and lipases (from *Rhizopus oryzae, Chromobacterium viscosum, Fusarum solanii* (Cutinase)) were used as reference systems in order to establish combinations of different fluorogenic substrates (glycerolipids and single-chain carboxylic acid esters) for the discrimination of the lipolytic enzymes in biological samples. Using enantiopure substrates, we identified three different lipolytic activities (two lipases and one esterase) in lysates of *Bacillus subtilis* cells (Hermetter *et al.*, 1998). In addition, lipase and/or esterase activities as well as stereoselectivities of two new lipolytic enzymes (RR1 and RR2) and other esterases were characterized in different solvents.

2.4 Structure-function relationships

Enantiopure 1(3)-*O*-alkyl-2,3(1,2)-diacylglycerols (Figure 1, compounds B) were applied as substrates in order to obtain information on structure-function relationships of microbial and animal lipases. Two studies will be reported here, namely on a fungal enzyme (from *Rhizopus oryzae*) and human hepatic lipase. Mutations in the close vicinity of the catalytic triad of recombinant lipases from *Rhizopus oryzae* (expressed from the corresponding plasmid DNA constructs in E.coli) gave rise to dramatic changes in activity and stereoselectivity towards the long-chain alkyldiacylglycerols (A. Hermetter, E. Zenzmaier, F. Paltauf, unpublished) (Figure 2). Replacement of Leu 258 by the smaller Ala or Ser, but not by the larger Phe, led to a significant loss of stereoselectivity, most likely as a consequence of larger motional freedom of the lipid in the active site. The same mutations in the enzyme did not affect stereoselectivity towards the short-chain substrate which was *a priori* low. Binding of the latter compound to the protein might be looser as compared to the long-chain analogue. As a consequence, these interactions are likely not to be influenced by small steric changes in the protein binding site. Replacement of Ala 88 in the lid by Trp did not show any effect (Figure 2). Obviously this position of the lid does not play any role in stereospecific substrate recognition. Recombinant human hepatic lipases were expressed in Cos 7 cells transfected with the corresponding plasmid DNA constructs (M. Duque, H. Will and A. Hermetter, unpublished), including the wild-type enzyme and mutants amino-terminally truncated to a different extent. Enzyme activities were screened in the cell culture supernatants using long-chain alkyldiacylglycerol as a substrate.

The fluorescence assay was further used to screen a very large number of samples of recombinant lipases in order to determine enzyme activities depending on the structure of the amino-terminal domain accomodating the active site. Shortening of the amino terminus by nine amino acids led to almost complete inactivation. Thus, the first amino acids of this segment of the polypeptide chain seem to be essential for protein folding and function. According to a recent X-ray crystallographic study on cholesterol esterase,

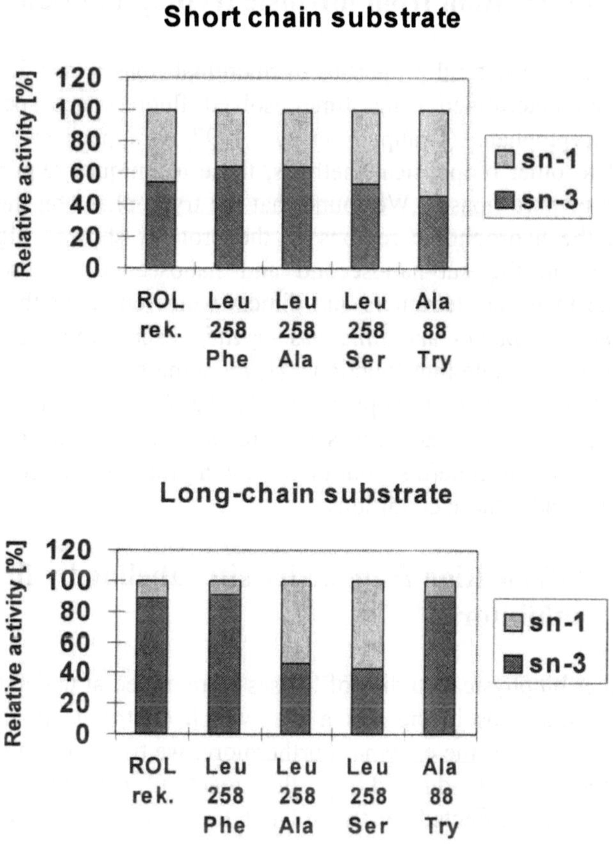

Figure 2. Stereoselectivity of a lipase from *Rhizopus oryzae*. Effect of amino acid exchange in the substrate binding site (Leu258) and the lid (Ala88). Substrates: 1(3)-O alkyl-2,3(1,2)-diacyl glycerols (see Figure1). Reaction medium: EtOH/H_2O (1/3, v/v) *sn*-1, *sn*-3: positions of the scissile acylester bond (TNP-acyl chain) in the substrate.

the carboxy-terminal segment seems to play a specific role in the activation of this enzyme (Chen *et al.*, 1997). It contains the active site and moves away upon interaction with the substrate. Since the three-dimensional structure of hepatic lipase is not available yet, we can only speculate about a similar role of the amino terminus in this enzyme.

3. PHYSICAL PROPERTIES OF LIPOLYTIC ENZYMES IN SOLUTION

3.1 Information from intrinsic tryptophan fluorescence

The gross structural properties of microbial lipases (see 2.3) in solution have been determined using time-resolved fluorescence spectroscopy of protein tryptophans (Graupner *et al.*, 1997, Graupner *et al.*, 1999). In contrast to other biophysical methods, these techniques require only small amounts of pure lipases. We found that the tryptophans on the surface and even in the hydrophobic regions of the proteins showed high segmental flexibility on the sub-nanosecond and nanosecond time-scale. Lipases aggregate in water (tetramers and dimers were found for the lipases from *Pseudomonas species* and *Rhizopus oryzae,* respectively), but the protein complexes dissociate into functional (!) monomers, when organic solvent or detergent is present (Graupner *et al.*, 1997, Graupner *et al.*, 1999). Preservation of lipase activity is in agreement with fluorescence emission spectra and decay time(distribution)s showing that the respective lipases do not unfold under these conditions.

3.2 Information from active site-labeling by fluorescent inhibitors

Further biophysical studies of lipases were aimed at understanding lipid-protein interactions in the active site, which contribute to the regio- and stereoselectivity of the enzyme. Furthermore, we tried to find out how these interactions are modulated by the reaction medium since enzyme selectivities are affected by the solvent to a great extent. Most oft these studies have been carried out with the lipase from *Rhizopus oryzae*. For this purpose, we have synthesized long- and short-chain glycerolipid inhibitors carrying a fluorescent reporter covalently bound to the ω-end of the hydrophobic chains in positions *sn*-1, -2, or -3 of glycerol (Figure 3). The respective compounds were dialkylglycero-alkylphosphonic acid-*p*-nitrophenyl esters which react with the nucleophilic serine in the active site, and lead to a covalent complex mimicking the initial transition state during substrate-lipase interaction (Figure 3). Using HPLC we isolated the four possible stereoisomers of each compound, resulting from the presence of two optical centers (glycerol-C2 and P) of the glycerolipids and determined the kinetics of enzyme inhibition by the respective lipids (Zandonella *et al.*, 1999). The tentatively assigned $S_C S_P$ and $R_C S_P$ isomers (based on the

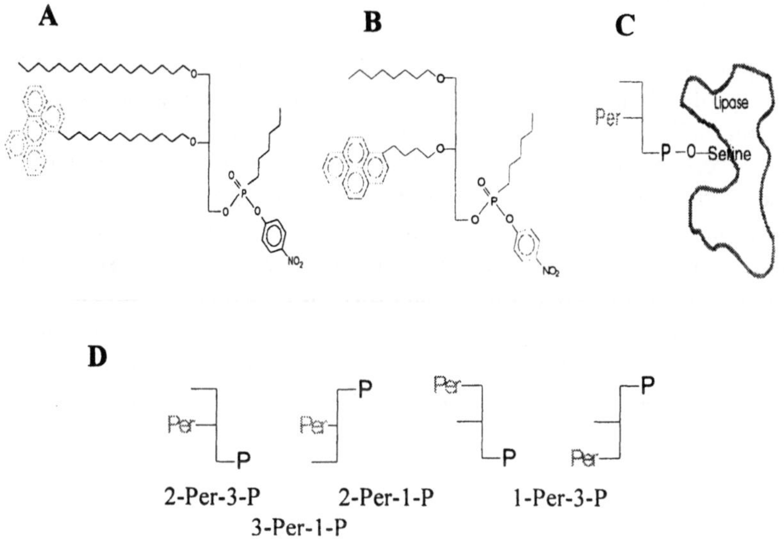

Figure 3. Fluorescent inhibitors for lipolytic enzymes. A: 1-*O*-Hexadecyl-2-*O*-perylenedodecyl-*sn*-glycero-3-hexyl-phosphonic acid-*p*- nitrophenyl ester. B: 1-*O*-Octyl-2-*O*-pyrenebutyl-*sn*-glycero-3-hexyl-phosphonic acid-*p*- nitrophenyl ester. C: Inhibitor A covalently linked to the nucleophilic serine of a lipase. D: Inhibitor isomers fluorescently labeled with perylene (PER), at different positions of glycerol.

assumption that the S_P enantiomer is much more reactive (Stadler *et al.*, 1996)), which turned out to be the most efficient inhibitors, were used for the biophysical studies. Using the respective compounds, lipase-inhibitor complexes were obtained containing 1 mol lipid per 1 mol protein.

Two classes of inhibitors were synthesized depending on the type of fluorophore. Pyrene represents a distance-dependent energy acceptor for protein tryptophans and thus monitors conformational changes of the protein in the lipid-bound state. Perylene on the other hand, provides information on the mobility of the labeled lipid segment in the presence of the protein. Recent X-ray crystallography data showed that only the first four methylene groups of the inhibitor fatty acids interact with the lipase from *Pseudomonas cepacia* (Lang *et al.*, 1998). The ω-end of the hydrophobic chain is not in intimate contact with the protein, although its mobility is influenced by the degree of immobilization of the carboxy-end. Thus, a fluorophore in the ω-position of a hydrophobic side chain can provide useful information about binding of the labelled lipid segment to the lipase.

3.3 Effect of lipid structure and configuration on lipid-protein interactions

Regioselective interactions of the hydrophobic lipid segments with the catalytic site of the lipase from *Rhizopus oryzae* were studied using three different dialkylglycerophosphonates labeled with perylene at the *sn*-1, -2, -and -3 alkyl residues, respectively (Figure 3). Equimolar complexes of the lipase with the individual inhibitors were prepared and fluorescence anisotropies were determined as a measure of (lipid) label rigidity. In lipase-bound form, the labeled *sn*-2 chain of the lipid showed the highest rigidity (r = 0.15), whereas the labeled *sn*-1 chain was more flexible (r = 0.13).

The mobility of the labeled *sn*-3 chain (r = 0.12) remained almost unchanged as compared to the inhibitor in a protein-free lipid (micelle) system (Figure 4). Thus, we suppose, that the *sn*-2 alkyl chain is responsible

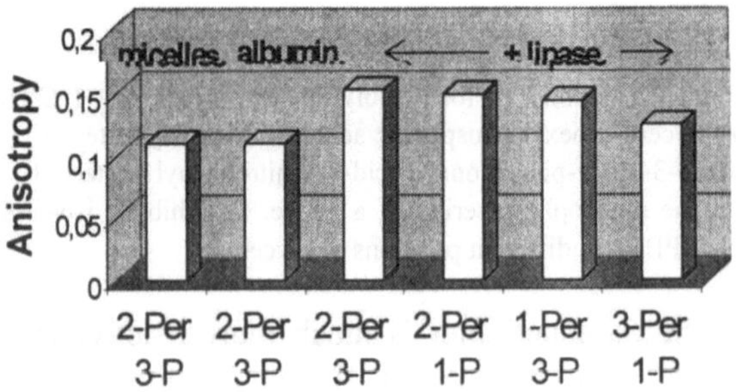

Figure 4. Regioselective inhibitor-protein interaction. Fluorescence anisotropies were determined according to $r = I_V - I_H / I_V + 2I_H$, I_V and I_H being the fluorescence intensities parallel and vertical to the plane of the vertically polarized excitation light. Anisotropy and thus, rigidity of the *sn*-2 labeled alkyl chain is highest in the lipase-bound inhibitor as compared to the glycerolipid in micelles or in the presence of albumin, indicating immobilisation of the lipid by the enzyme. In contrast, rigidity of the labeled *sn*-3 alkyl chain is unaffected by the lipase.

for strong lipid binding to the enzyme, whereas the *sn*-3 alkyl residue may form a bridge to the supramolecular substrate (Figure 5). The configuration of glycerol-C2 did not strongly affect lipid-protein interaction as determined with glycerolipid inhibitors fluorescently labeled with pyrene or perylene in position *sn*-2 Thus, the *sn*-2-alkyl residue seems to be accomodated in the

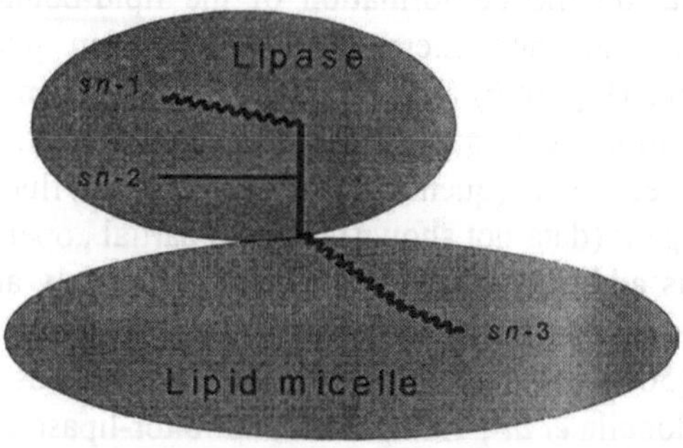

Figure 5. Model of lipid-protein interaction for the lipase from *Rhizopus oryzae*. The *sn*-2 alkyl chain of the glycerolipid is strongly immobilized by the enzyme. Interaction of the lipase with the *sn*-1 alkyl chain is weaker. Only very little interaction occurs between the lipase and the *sn*-3 alkyl chain. It may function as a bridge between the lipase and the lipid particle.

same binding pocket of the lipase, irrespective of glycerol stereochemistry (R_C or R_S). The above results are likely to be the first experimental data for lipases in solution, supporting a current hypothesis which is based on molecular modeling studies of physical interactions between triacylglycerols and the lipase from *Rhizopus oryzae*. It has been proposed that the *sn*-2 alkyl chain is tightly bound, whereas the hydrophobic residue in position *sn*-3 is more flexible (Scheib *et al.*, in press).

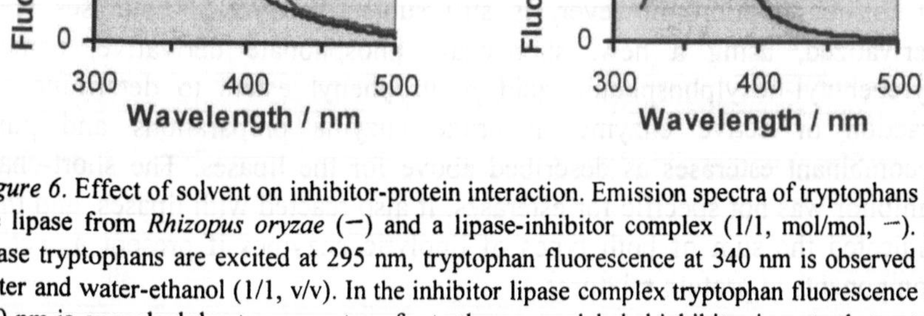

Figure 6. Effect of solvent on inhibitor-protein interaction. Emission spectra of tryptophans in the lipase from *Rhizopus oryzae* (—) and a lipase-inhibitor complex (1/1, mol/mol, —). If lipase tryptophans are excited at 295 nm, tryptophan fluorescence at 340 nm is observed in water and water-ethanol (1/1, v/v). In the inhibitor lipase complex tryptophan fluorescence at 340 nm is quenched due to energy transfer to the pyrene-labeled inhibitor in water but not in water-ethanol. The inhibitor was compound A in Figure 3.

Solvent effects on the conformation of the lipid-bound protein were determined from resonance energy transfer between pyrene-lipids and protein tryptophans (Figure 6) (Zandonella *et al.*, 1999). In aqueous buffer, the lipid-lipase complex seems to adopt a closed conformation, since energy transfer is very efficient (quenching of tryptophan fluorescence). The presence of detergent (data not shown) leads to partial „opening" of enzyme structure, whereas addition of organic solvent completely abolishes energy transfer and thus gives rise to extensive unfolding. Only at high solvent concentrations (50%) the enzyme completely unfolds and becomes inactivated (Zandonella *et al.*, 1995). Thus, inhibitor-lipase conjugates could be useful tools for a simple spectroscopic assay to find out whether or to what extent the activity of a given lipolytic enzyme is affected by the reaction medium.

4. FLUORESCENT LIPASE INHIBITORS AS ANALYTICAL TOOLS

Fluorescent inhibitors are sensitive and specific reagents for detection and quantification of lipolytic enzymes in complex biological samples. We found that long-chain dialkylglycerophosphonates selectively reacted with a microbial lipase in pure form (from *Rhizopus oryzae*) or a lipase from *Rhizomucor miehei* in a commercial enzyme preparation. Separation of the individual protein components by SDS electrophoresis showed in both cases a single fluorescent band corresponding to the molecular weights of the corresponding enzymes (Results for the *Rhizomucor lipase* are shown in Figure 7). The fluorescence intensity patterns of the gels were determined after electrophoresis, using a fluorescence plate reader or a CCD-camera. The obtained values linearly depended on protein mass and, since the inhibitor reaction is stoichiometric, provided a measure for the fraction of active enzyme in the protein sample. The long-chain glycerolipid did not react with esterases and seems to be specific for lipases (the general validity of this assumption, however, is still subject to proof). Esterases were derivatized, using a new, short-chain phosphonate derivative, namely pyrenebutyl-hexylphosphonic acid *p*-nitrophenyl esters to determine the fraction of active enzyme in crude enzyme preparations and pure, recombinant esterases as described above for the lipases. The short-chain inhibitor was not specific for esterases. It also reacted with lipases, and thus indicated the sum of both types of lipolytic enzymes if present as active components in protein mixtures.

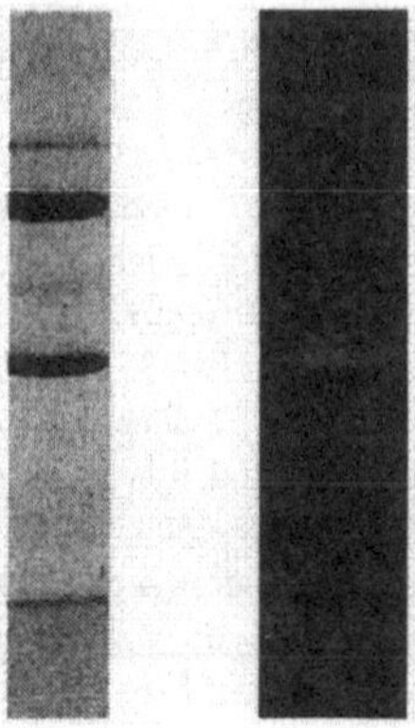

Figure 7. Lipase imaging on polyacrylamide gels. SDS-gel electrophoresis of a crude lipase from *Rhizomucor miehei* preincubated with inhibitor 2-PER-3P (compound A, Figure 3) solubilized by Triton X-100 (2.5 mM) in aqueous buffer (0.1 M Tris/HCl pH 7.4). Incubation was carried out at 30°C for 3 hours. **1** protein stained with Coomassie blue; **2** fluorescent lipid-protein complex under UV light excitation at 350 nm.

The application of fluorescent alkylglycerophosphonates as inhibitors of lipolytic enzymes seem to be very promising in research, industry, and medical diagnosis. In particular, combinations of inhibitors labeled with different dyes and specific for different types of enzymes appear to be attractive for (automated) multiwavelength analysis of lipase/esterase mixtures and thus, open new perspectives for enzyme discrimination.

ACKNOWLEDGMENT

We acknowledge financial support by the Austrian research foundation (FWF-Project F0107).

REFERENCES

Anthonsen H. W., Baptista A., Drablos F., Martel P., and Petersen S., 1994, The blind watchmaker and rational protein engineering. *J. Biotech.* **36**: 185-220.

Anthonsen M. W., Baptista A., Orablos F., Matel P., Petersen S. B., Sebastiao M., and Vaz L., 1995, Lipases and Esterases: a review of their sequences, structure and evolution. *Biotech.Annu. Rew.* (M. R. El-Gewely, ed.) pp. 319-371.

Chen J. C.-H., Miercke L. J. W., Krucinski J., Starr J. R., Saenz G., Wang X.,

Duque M., Graupner M., Stütz H., Wicher I., Zechner R., Paltauf F., and Hermetter A., 1996, New fluorogenic triacylglycerol analogs as substrates for the determination and chiral discrimination of lipase activities. *J. Lipid. Res.* **37**: 868-876.

Faber K., 1992, In *Biotransformation in Organic Chemistry* Springer, Berlin Heidelberg.

Graupner M., Haalck L., Spener F., Paltauf F., and Hermetter A., 1997, Molecular dynamics of microbial lipases in different solubilization systems as determined phase-resolved fluorometry. In *Ultrafast Spectroscopy*.(O. Svelto, S. de Silvestri, and G. Denardo, eds.) Plenum Press, New York, pp. 531 - 533.

Graupner M., Haalck L., Spener F., Lindner H., Glatter O., Paltauf F. and Hermetter A., 1999, Molecular dynamics of microbial lipases as determined from their intrinsic tryptophan fluorescence. *Biophys.J.* **77**: 493 – 504.

Hermetter A., Paltauf F., Zenzmaier E., Schlacher A., Schwab H., and Faber K., 1998, Fluorescence screening for lipolytic enzymes. In *New Frontiers for Microbial Catalysts* (K. Kieslich, C. P. Van der Beck, J. A. M. De Bout, and W. J. J. Van den Tureel, eds), Elsevier, Amsterdam, pp. 53-60.

Hermetter A., 1998, Lipase assay based on a novel fluorogenic alkyldiacylglycerol substrate. In *Methods in Molecular Biology*: Lipase and Phospholipase Protocols (M.H. Doolittle and K. Reue, eds.), Humana Press, Inc., Totowa, NJ, **109**: 19-29.

Lakowicz J. R., 1986, Principles of fluorescence spectroscopy, Plenum Press, New York and London.

Lang D.A., Mannesse M. L. M., de Haas G.H., Verheij M. M., and Dijkstra B. W., 1998, Structural basis of the chiral selectivity of *Pseudomonas cepacia* lipase. *Eur. J.Biochem.* **254**: 333-340.

Scheib H., Pleiss J., Kovac A., Paltauf F., and Schmid K. D., Steroselectivity of Mucorale lipases towards triradylglycerols – a simple solution to a complex problem. *Protein Sci.*, in press.

Schmid R.D. and Verger R., 1998, Lipasen: Grenzflächenenzyme mit attraktiven Anwendungen. *Angew. Chemie* **110**: 1694-1720.

Scholze H., Stütz H., Paltauf F. and Hermetter A., 1999, Fluorescent inhibitors for the qualitative and quantitative analysis of lipolytic enzymes. *Anal. Biochem.* **276**: 72 - 80

Spilburg C. A., Lange L. G., Ellsworth J. L., and Stroud R.M., 1997, Structure of Bovine Pancreatic Cholesterol Esterase at 1.6 Å: Novel Structural Features Involved in Lipase Activation. *Biochemistry* **37**: 5107-5117.

Stadler P., Zandonella G., Haalck L., Spener F., Hermetter A., and Paltauf F., 1996, Inhibition of microbial lipases with stereoisomeric triradylglycerol analog phosphonates. *Biochim. Biophys. Acta* **1304**: 229-244.

Verger R., 1997, „Interfacial Activation“ of Lipases: Facts and Artifacts.*TIBTECH* **15**: 32-38.

Zanonella G., Haalck L., Spener F., Faber K., Paltauf F.and Hermetter A., 1995, Inversion of Lipase Stereospecifity for Fluorogenic Alkyldiacyl Glycerols. Effect of Substrate Solubilization. *Eur. J. Biochem.* **231**: 50-55.

Zandonella G., Stadler P., Haalck L., Spener F., Paltauf F. and Hermetter A., 1999, Interaction of fluorescent triacylglycerol analogs covalently bound to the active site of a lipase from Rhizopus oryzae. *Eur.J.Biochem.* **262**: 63 - 69

Infrared Spectroscopy of Lipoproteins

XABIER COTO AND JOSÉ LUIS R. ARRONDO
Unidad de Biofísica and Departamento de Bioquímica, Universidad del País Vasco, Apdo. 644, E-48080 Bilbao, Spain

1. INTRODUCTION

Lipids are transported in human blood plasma by lipoproteins consisting of a nonpolar core where triacylglycerols and cholesteryl esters are hidden surrounded by a monolayer facing the water composed of phospholipid, cholesterol and proteins, giving these lipid-rich structures water solubility. Blood plasma lipoproteins are classified on the basis of their density, which in turn is a reflection of their lipid content. The greater their lipid contents the lower their density. Three major lipoprotein classes are found in fasting human blood plasma, VLDL or very low density lipoprotein, LDL or low density lipoprotein and HDL or high density lipoprotein. Their characteristics are depicted in Table 1.

Table 1. Composition of lipoproteins from human blood serum in fasting subjects

	VLDL	LDL	HDL
Density	0.95-1.006	1.019-1.063	1.063-1.210
Electrotoretic	Pre-β	β	α
Half-life	1-3 h	2.5-3.5 days	5-6 days
Proteins (%)	B-100, C-I, C-II, C-III, E (8)	B-100 (22)	A-I, A-II (40-50)
Triglycerides	55	6	8
Phospholipids	18	22	22
Cholesterol	7	8	3-5
Cholesteryl esters	12	42	14-17

Biosynthesis of VLDL cholesterol, triglycerides and apolipoproteins takes part in liver hepatocites. These particles have apoB-100, apoC and apoE proteins. VLDL are converted into LDL, cholesteryl ester-rich particles

Supramolecular Structure and Function 7, edited by Pifat-Mrzljak.
Kluwer Academic/Plenum Publishers, New York, 2001.

that have a single molecule of apoB-100. LDL carries cholesterol from liver to peripheral tissues where it is used in membrane and steroid biosynthesis. HDL is synthesised from liver precursors and are matured in plasma. Apolipoproteins in HDL are apoA-I and apoA-II. HDL removes cholesterol from the cells carrying it to the liver in a “reverse transport”.

Oxidative modification of LDL (oxLDL) has been implicated in the formation of lipid-laden cells, which is an early step in the development of atherosclerosis (Ross, 1993). Once formed by the endothelium, oxLDL play an initial role in the sequence of events leading to the activation of the monocytes into macrophages. Uptake of oxLDL by the macrophages will lead to foam cell formation and may alter gene expression of many of the growth-regulatory hormones, cytokines and substances of low molecular mass involved in plaque formation.

Infrared spectroscopy is nowadays a useful technique in the study of protein conformation and dynamics. The possibilities of the technique become apparent specially when applied to large proteins in turbid suspensions, as is the case with lipoproteins. Variations in lipoprotein structure have been probed by monitoring the IR Amide I band. This band arises mainly from C=O stretching vibrations and is sensitive to conformational changes. Infrared spectroscopy is a choice method, not only as an alternative option or as a complement to the high-resolution techniques in membrane protein studies, because the lipid environment not only does not perturb the resolution or sensitivity of the spectra but can also be the object of simultaneous study, thus providing information on the lipid-protein interactions. Even if infrared spectroscopy was first applied to proteins as early as in 1952 (Sutherland) before any detailed X-ray results were available, its use in proteins in general has only been possible after the revolution in instrumentation produced by the development of microcomputers that allowed the design of instruments based on the Michelson interferometer and the fast Fourier transform. In principle, a structure as large as a protein would give rise to an enormous number of overlapping vibrational modes obscuring the information that could be obtained in practice, but because of the repeating patterns of the biological molecules, e.g. the secondary structure of the protein backbone, the spectra are much simpler and useful structural information can be obtained.

The structural analysis usually implies a mathematical approach in order to extract the information contained in the composite bands, designated in IR spectroscopy as 'amide bands', obtained from proteins. Commonly used methods of analysis imply narrowing the intrinsic bandwidths to visualize the overlapping band components and then decomposing the original band contour into these components by means of an iterative process. The various components are finally assigned to protein or subunit structural features (Arrondo *et al.*, 1993).

2. LOW DENSITY LIPOPROTEINS

LDL are the particles that have been studied more extensively by IR spectroscopy, not only because they are involved in cholesterol deposition but also because apoB, their apolipoprotein cannot be isolated and studied by other structural techniques. ApoB is one of the largest proteins known, containing 4536 aminoacid residues, and is extremely insoluble in aqueous media. This has hindered the understanding of the structural basis of apoB function, although its primary structure has been known for some time (Knott *et al.*, 1986, Yang *et al.*, 1986). IR was first applied to the study of apoB by Scanu *et al.*, (1969) to study its thermal behaviour. Later IR was applied to apoB studies qualitatively using resolution-enhanced techniques (Herzyk *et al.*, 1987), or quantitatively using curve fitting of deconvolved spectra (Goormaghtigh *et al.*, 1989). Also, IR has described the structure of the lipid-attached protein in combination with proteolytic digestion (Goormaghtigh *et al.*, 1993).

Decomposition of the amide I band of human plasma LDL apoB in H_2O and D_2O media at physiological temperature is shown in Figure 1. The corresponding parameters to the H_2O spectrum, *i.e.* band position, percentage area and bandwidth of each spectral component are displayed in Table 2. The D_2O spectrum exhibits seven bands assignable to protein substructures located at 1694, 1678, 1670, 1656, 1643, 1630 and 1618 cm^{-1}. In the H_2O spectrum, six bands located at 1695, 1683, 1670, 1653, 1632 and 1618 cm^{-1} are seen.

The assignment of these bands has been carried out previously (Bañuelos *et al.*, 1995). The band at 1656 cm^{-1} in D_2O is undoubtedly assigned to α-helix, and the one around 1630 cm^{-1} to ß-sheet. The bands at 1670 and 1680 cm^{-1} arise from ß-turns. The band at 1643 cm^{-1} in D_2O corresponds to unordered structure; in H_2O this band is shifted to around 1657 cm^{-1} and overlaps with the α-helix structure. It is not usual to find a protein band at frequencies as low as 1618 cm^{-1}. In fact, in non-denatured proteins in H_2O solution, it has only been found in apolipoproteins and assigned to 'low frequency ß-sheets' (Goormaghtigh *et al.*, 1989) or to a ß-structure less accessible to the external solution (Herzyk *et al.*, 1987). According to its position, this band must be assigned to an extended structure, even more so considering the presence of a band at 1694 cm^{-1} that corresponds to the *(0,B)* vibrational mode of the extended chain. However, it cannot be assigned to antiparallel ß-sheet because the high frequency component does

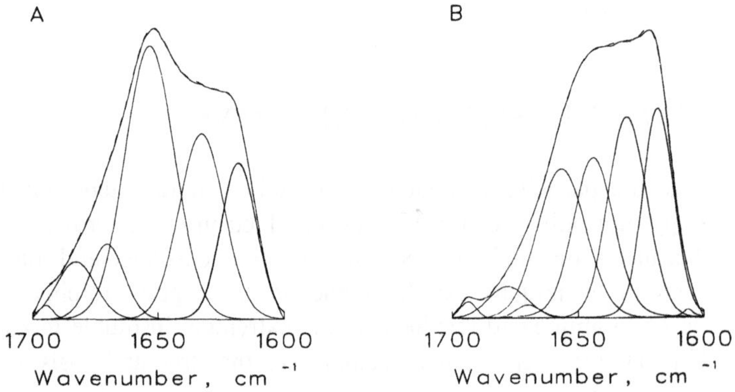

Figure 1. Amide I band decomposition of apoB from human LDL at 37 °C in H_2O (A) and D_2O (B) in 0.1 M Tris, pH or pD 7.4 buffer. The parameters corresponding to the component bands are reflected in Table 2.

Table 2. Amide I band decomposition of LDL amide at 37 °C in H_2O medium.

Band position	% Band area	Bandwidth
1695	7	5
1683	5	18
1670	5	15
1653	45	24
1632	25	22
1618	19	16

The figures have been rounded off to the nearest integer

not undergo the isotopic shift typically produced in that kind of structure by changing the medium from H_2O to D_2O (Susi, 1969). A more generalised assignment would be to ß-strands, i.e. extended structures not forming ß-sheet. From the combined results in H_2O and D_2O, the secondary structure of apoB would be 24% α-helix, 23% ß-sheet, 4% ß-turns, 24% unordered structure, and 24% ß-strands. Information on the apoB attachment to the lipid surface can be obtained by studying the unique characteristics of some LDL protein bands. The most striking feature of the Amide I band is the component at 1618 cm^{-1}. Its thermal behaviour is different from that of other components, and follows a pattern similar to the methylene stretching of the lipid moiety (Figure 2).

This would support an assignment of this band component as a protein structure penetrating the monolayer and establishing hydrophobic interactions deeper than the outer charged shell. The existence of protein regions inside the monolayer has been deduced from trypsin controlled

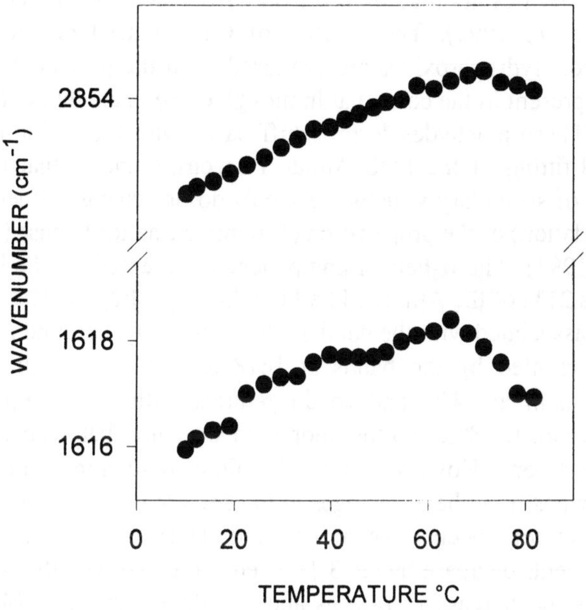

Figure 2. Band position *versus* temperature plot for two LDL infrared spectral features. Band positions correspond to the lipid CH_2 symmetric stretching (≈ 2854 cm^{-1}), and to the protein ß-strand conformation (≈1618 cm^{-1}).

proteolysis that shows non-accessible regions of the protein molecule tightly associated with LDL (Goormaghtigh *et al.*, 1993). The assignment of the 1618 cm^{-1} band to this protein fraction interacting with the lipids is also compatible with electron microscopy studies showing a component embedded in the lipid and connecting the different apoB-100 protein domains (Spin and Atkinson, 1995). The increased strength of hydrogen bonds within the hydrophobic milieu would account for the 7 cm^{-1} difference existing between the position of this component in lipoprotein and in concanavalin A (1625 cm^{-1}), or cytochrome C oxidase (1624 cm^{-1}). The amount of extended chain interacting with the lipid moiety (18-24 %, see Table 2) would involve between 820 and 1090 peptide bonds. Assuming a displacement of 3,47 Å per residue, the length of the chain would wrap more than four times the equator of the LDL particle. Therefore, these results would match a disposition of the protein as a long flexible string as postulated by immunoelectron microscopy (Chatterton *et al.*, 1991).

Oxidatively modified LDL seems to play a significant role in the initiation and progression of atherosclerosis *in vivo* (Ross, 1993, Ylä-Herttuala, 1991). Metal ions such as Cu^{2+} or Fe^{2+} are frequently used to initiate LDL oxidation *in vitro*, since the resulting oxLDL has similar biological activities to that oxidized *in vivo* (Steinbrecher *et al.*, 1984, Estebauer *et al.*, 1992). The kinetics of Cu^{2+}-induced oxidation is well characterized. Hydroperoxides are produced from the polyunsaturated fatty acyl chains present in the core, that in turn give rise to aldehydes (Pinchuk *et al.*, 1998). These aldehydes form Schiff bases with the lysine residues of apoB. Band-fitting of the LDL Amide I spectrum shows that the relative proportions of secondary structure of apoB do not change either in the lag (15 min oxidation) or the propagation (110 min oxidation) phase (Gieseg and Estebauer, 1994). The α-helical component represented by the band at *ca.* 1656 cm^{-1} is 21% of the Amide I in all the three spectra, whereas the *β*-sheet component associated with the band at 1631 cm^{-1} is also around 21%. The *β*-strands represented by the bands at 1618 and 1694 cm^{-1} in the control spectrum remain at 24%, and so do *β*-turns, with two components that together account for 8%, and the unordered structure that accounts for 27% of theAmide I band. However, after 180 min of oxidation, at the beginning of the decomposition phase, changes in the percent distribution of the apo B Amide I band components can be detected. Thus, the *α*-helix and *β*-sheet band components decrease by *ca.* 3 % when compared with the control, with a corresponding increase in *β*-turns and unordered structure. Moreover, in the case of *β*-turns, of the two bands associated to this structure, the component around 1680 cm^{-1} is lost and the one at 1670 cm^{-1} increases from around 5% up to 12%, indicating a reorganisation in the *β*-turns. It must be noted that the *β*-strand component, attributed to extended chains penetrating the monolayer, does not vary upon oxidation at least before the decomposition phase. However, in LDL particles subjected to very long oxidative periods, this band is also affected. Looking at the effect of temperature on the protein in its native state or after modification by oxidation can also monitor changes in apoprotein B conformation. The significance of studying temperature-induced protein conformational changes by infrared spectroscopy has been previously reported for membrane and soluble proteins, as well as for lipoproteins (Chehin *et al.*, 1998). In Figure 3 the band positions obtained from the deconvolved spectra of control and oxidized LDL, and corresponding to *α*-helix (.1654 cm^{-1}), *β*-sheet (.1630 cm^{-1}) and the *β*-strands (.1617 cm^{-1}), have been plotted against temperature using a continuous heating method.

In all three bands, the profiles corresponding to control and 15 min-oxidized LDL particles are virtually indistinguishable. However, after 180 min oxidation, when a maximum concentration of dienes is reached, the temperature profile corresponding to the *α*-helix and *β*-sheet has been displaced to lower wavenumbers, indicating variations in tertiary but not in

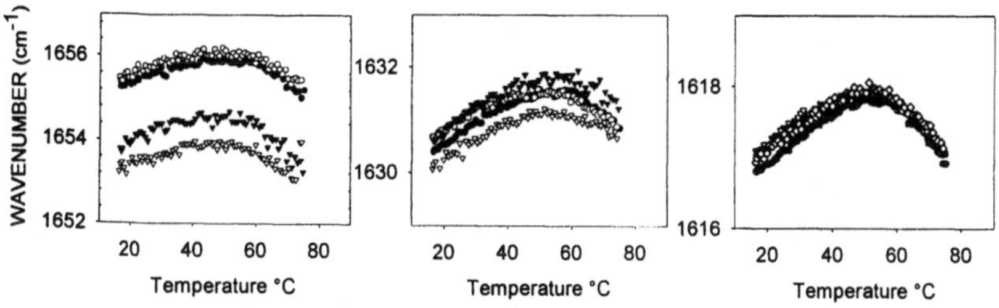

Figure 3. Thermal profiles of the bands corresponding to α-helix (top), β-sheet (middle) and β-strands (bottom) obtained from the amide I band of apoB from human LDL and corresponding to 0 (•), 15 (○), 110 (▼) and 180 (▽) minutes of oxidation.

secondary structure (Arrondo and Goñi, 1999). The band at 1617 cm^{-1} that is attributed to structures embedded in the lipid monolayer is not affected after 180 min oxidation. At 110 min oxidation, the situation is intermediate between the two described above, i.e. the temperature profile of the 1630 cm^{-1} band (*β*-sheet) is not affected by oxidation, whereas the band corresponding to *α*-helix at 1654 cm^{-1} is displaced almost as much as after 180 min oxidation. The results presented above show that even in the early oxidation stages of LDL, in the absence of changes in the secondary structure or physical state of the particle, apoB undergoes changes in tertiary structure, primarily in the *α*-helical and then in the *β*-structure component, that could be associated with changes on the particle surface. Such an altered surface would then promote the binding of LDL to putative receptors and/or accumulation of cholesterol.

3. VERY LOW DENSITY LIPOPROTEINS

VLDL is secreted to plasma from liver and is enriched in triglycerides when compared with LDL. More proteins are present in the outer layer, but among them apoB-100 are pre-eminent. The Amide I band fitting and the percentage of the components are shown in Table 3.

Amide I shape is dominated by the bands corresponding to the β-strands arising from apoB-100, indicating so the predominance of this protein in the particle apolipoproteins. Assignment of the infrared components can be made parallel to LDL, and from Table 3, the secondary structure of VLDL would be 28 %α-helix, 23 % β-sheet, 5% β-turns, 20% unordered structure and 23% β-strands. ApoB-100 prevalence in the protein particle structure is also reflected in this secondary structure, which is close to LDL.

Table3. Amide I band decomposition of the VLDL particles.

Band position	% Band area	Bandwidth
1694	<1	6
1683	<1	16
1670	5	18
1656	28	19
1644	20	17
1632	23	20
1618	23	16

The three dimensional profile of VLDL thermal denaturation (Figure 4) is slightly different from a classical protein profile (Chehin *et al.*, 1998, Arrondo and Goñi, 1999) because the low frequency band corresponding to aggregation is very close to the one arising from the β-strands. However, this does not mean that the bands attributed to β-strands have structural characteristics of an aggregated protein, since the high frequency component, located at 1693 cm^{-1} in the β-strands and not affected by the isotopic shift produced when the protein is measured in a D_2O medium (Susi, 1969), disappears after protein denaturation.

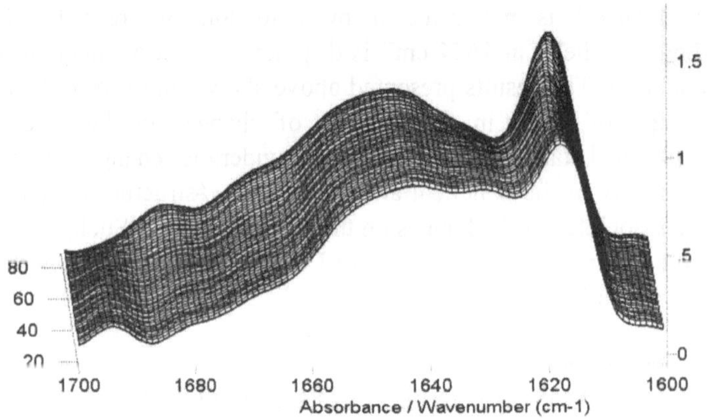

Figure 4. 3D plot of VLDL thermal denaturation in the amide I region.

A more detailed picture of protein denaturation and the lipid-protein connection can be seen in Figure 5 where the band corresponding to β-strands are plotted together with a band coming from the cholesteryl ester C=O stretching (≈ 1738 cm^{-1}) and another from the CH_2 stretching from the acyl chains (≈ 2852 cm^{-1}). From the Figure 5 it is clear that even if in the lipid moiety a phase transition is not seen, changes in the protein embedded in the lipid monolayer affects even the acyl chains located in the lipoprotein core, although to a lesser extent than in LDL (Bañuelos *et al.*, 1995),

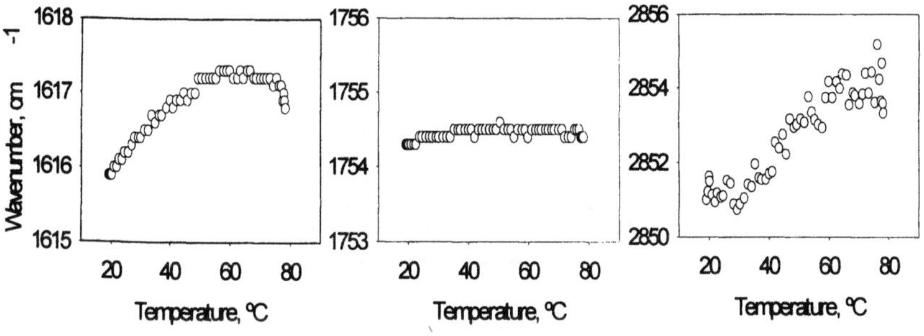

Figure 5. Wavenumber *versus* temperature plots of the infrared bands corresponding to the β-strand of VLDL apolipoprotein (left), the carbonyl bands of the lipid moiety (centre) and the CH_2 stretching bands of the acyl chains (right).

what could also be expected by the different size and composition of the particles.

4. HIGH DENSITY LIPOPROTEINS

HDL particles have different physiological functions than VLDL or LDL participating in the so-called "reverse cholesterol transport", which includes stimulation of cholesterol efflux from peripheral tissues and uptake of the particles by the liver *via* a specific receptor. This process may account for the inverse relationship observed between plasma high density lipoprotein concentration and the risk of developing coronary artery disease (Miller 1989). Also, this different function determines a different protein composition of the particle. Subsequently, the shape of the Amide I band is also different to LDL or VLDL since no apoB-100 is present (Figure 6).

A similar analysis to the other lipoprotein particles would give a secondary structure of 40% α-helix, which is similar to values observed previously by IR (Yang *et al.*, 1991). Also, a 27% β-sheet, 25% unordered structure and 8% β-turns are obtained. It must be noted the absence of the bands at 1618 and 1693 cm^{-1} characteristics of apoB-100 and associated with the β-strands embedded in the lipid. Also, HDL proteins can be isolated from the particle and reconstituted with lipids, what is not possible for apoB-100. The 3D-thermal denaturation plot (Figure 7) is similar to other proteins with a denaturation temperature in the range 60-70 °C, and showing the appearance of bands due to aggregation.

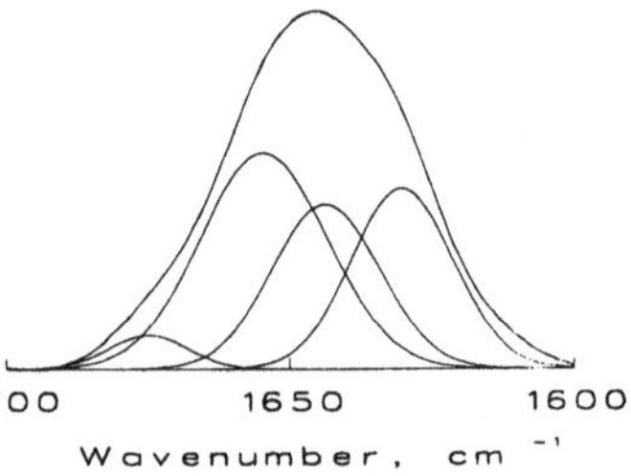

Figure 6. Amide I band decomposition of a HDL particle at 37 °C.

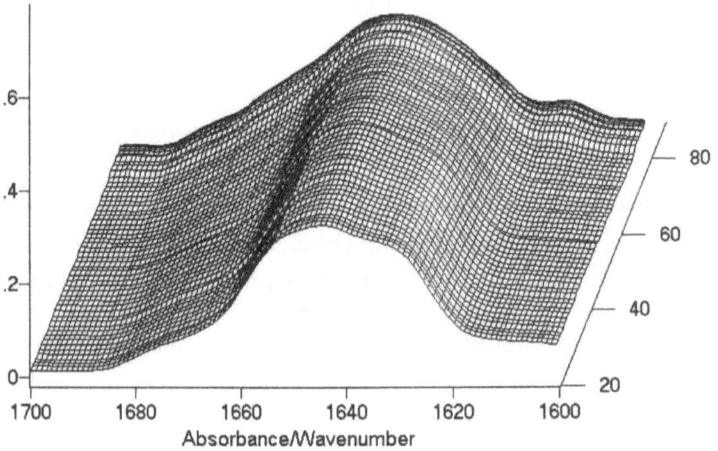

Figure 7. 3D plot of the Amide I thermal denaturation of HDL.

The major protein component of human HDL is apolipoprotein A-I (apohA-1) comprising a 70% of the total protein. From sequence analysis (Segrest *et al.*, 1992) and the crystal structure of a lipid-free truncation (apoΔ(1-43)A-I) (Borhani *et al.*, 1997) it is suggested that apohA-I forms a series of amphipatic α-helices. However, two disparate models best describe the orientation of these helices within the lipoprotein complex. A combination of attenuated total reflectance infrared spectroscopy (ATR) with circular dichroism, fluorescence energy transfer and a computer model (Brasseur *et al.*, 1990) gave rise to the so-called "picket-fence" model where six of seven helical repeats identified in apohA-I are located along the edges of a discoidal complex, with the axes of the helices oriented parallel to the phospholipid acyl chains. However, a more recent work using polarized attenuated total internal reflection Fourier-transform (PATIR-FTIR) spectroscopy (Koppaka *et al.,* 1999) supports the "belt" model, which also positions the helices of apohA-I around the perimeter of a lipid bilayer disc, but orients them parallel to the plane of the bilayer. This approach combines internal reflection infrared spectroscopy and Langmuir film balance technology and provides the information about lipid-protein interactions under conditions that are unlikely to disrupt the complexes or denature the protein components.

5. CONCLUSION

The distinct functionality of lipoproteins gives rise to differences in structure that are reflected in the infrared spectra. Thus, from the results presented above is clear that VLDL and LDL Amide I spectra are dominated by the special characteristics of apoB-100. The composition of the particle and the relationship core-surface is also reflected in the bands arising from proteins and lipids. The advantage of IR in the study of turbid suspensions and the variety of sampling techniques that can be used with lipoproteins prove the suitability of this technique in the structural studies of lipoproteins.

ACKNOWLEDGMENTS

This work has been supported by grants from University of the Basque Country (G03/98), DGICYT (PB96-0171), Basque Government (Ex1999-5 and PI1998-33).

REFERENCES

Arrondo, J.L.R., Muga, A., Castresana, J. and Goñi, F.M., 1993, Quantitative studies of the structure of proteins in solution by Fourier-transform infrared spectroscopy. *Prog. Biophys .Mol. Biol.* **59**: 23-56.

Arrondo, J.L.R. and Goñi, F.M., 1999, Structure and dynamics of membrane proteins as studied by infrared spectroscopy. *Prog. Biophys. Mol. Biol.* **72**: 367-405.

Bañuelos, S., Arrondo, J.L.R., Goñi, F.M., and Pifat, G., 1995, Surface-Core Relationships in Human Low Density Lipoprotein as Studied by Infrared Spectroscopy. *J. Biol. Chem.* **270**: 9192-9196.

Borhani, D.W.; Rogers, D.P.; Engler, J.A.; Brouillette, C.G., 1997, Crystal structure of truncated human apolipoprotein A-I suggests a lipid-bound conformation. *Proc. Natl. Acad. Sci. USA* **94**: 12291-12296.

Brasseur, R., De Meutter, J., Vanloo, B., Goormaghtigh, E., Ruysschaert, J.M., and Rosseneu, M., 1990, Mode of assembly of amphipathic helical segments in model high-density lipoproteins. *Biochimica and Biophysica Acta* **1043**: 245-252.

Chatterton, J.E., Phillips, M.L., Curtiss, L.K., Milne, R.W., Marcel, Y.L., and Schumaker, V.N., 1991, Mapping Apolipoprotein B on the Low Density Lipoprotein Surface by Immunoelectron Microscopy. *J. Biol. Chem.* **266**: 5955-5962.

Chehin, R., Iloro, I. And Arrondo, J.L.R., 1998, Thermal denaturation of membrane proteins: An infrared study. *Period. Biol.* **100**: Suppl2, 13-19.

Estebauer, H., Gebicki, J., Puhl, H., and Jürgens, G., 1992, Human plasma lipid peroxide levels show a strong transient increase after successful revascularization operations. *Free Rad. Biol. Med.* **13**: 341-390.

Gieseg, S.P. and Estebauer, H., 1994, Low density lipoprotein is saturable by pro-oxidant copper. *FEBS Lett.* **343**: 188-194.

Goormaghtigh, E., Cabiaux, V., De Meutter, J., Rosseneu, M., and Ruysschaert, J. M., 1993, Secondary structure of the particle associating domain of apolipoprotein B-100 in low-density lipoprotein by attenuated total reflection infrared spectroscopy. *Biochemistry* **32**: 6104-6110.

Goormaghtigh, E., De Meutter, J., Vanloo, B., Brasseur, R., Rosseneu, M., and Ruysschaert, J.M., 1989, Evaluation of the secondary structure of apo B-100 in low- density lipoprotein (LDL) by infrared spectroscopy. *Biochim. Biophys. Acta* **1006**: 147-150.

Herzyk, E., Lee, D.C., Dunn, R.C., Bruckdorfer, K.R., and Chapman, D., 1987, Changes in the secondary structure of apolipoprotein B-100 after Cu^{2+}-catalysed oxidation of human low-density lipoproteins monitored by Fourier transform infrared spectroscopy. *Biochim. Biophys. Acta* **922**: 145-154.

Knott, T.J., Pease, R.J., Powell, L.M., Wallis, S.C., Rall Jr, S.C., Innerarity, T.L., Blackhart, B., Taylor, W.H., Marcel, Y., Milne, R., Johnson, D., Fuller, M., Lusis, A.J., McCarthy, B.J., Mahley, R.W., Levy-Wilson, B., and Scott, J., 1986., Complete protein sequence and identification of structural domains of human apolipoprotein B. *Nature* **323**: 734-738.

Koppaka,V.; Silvestro,L.; Engler,J.A.; Brouillette,C.G.; Axelsen,P.H., 1999, The structure of human lipoprotein A-I. Evidence for the "belt" model. *J. Biol. Chem.* **274**: 14541-14544.

Miller, N.E., 1989, *High density lipoproteins, reverse cholesterol transport and coronary heart disease*. Elsevier. Amsterdan.

Pinchuk, I., Schnitzer, E., and Lichtenberg, D., 1998, Kinetic analysis of copper-induced peroxidation of LDL. *Biochim. Biophys. Acta* **1389**: 155-172.

Ross, R., 1993, The pathogenesis of atherosclerosis: a perspective for the 1990s. *Nature* **362**: 801-809.

Scanu, A., Pollard, H., Hirz, R., and Kothary, K., 1969, On the conformational instability of human serum low-density lipoproteins: Effect of temperature. *Proc. Natl. Acad. Sci. USA* **62**: 171-178.

Segrest J.P., Jones M.K., De Loof H., Brouillette C.G., Venkatachalapathi Y.V. and Anantharamaiah G.M., 1992, The amphipathic helix in the exchangeable apolipoproteins: a review of secondary structure and function. *J. Lipid Res.* **33**: 141-166.

Spin, J.M. and Atkinson, D., 1995, Cryoelectron microscopy of low density lipoprotein in vitreous ice. *Biophys. J.* **68**: 2115-2123.

Steinbrecher, U.P., Parthasarathy, S., Lee, D.C., Witztum, J.L., and Steinberg, D., 1984, Modification of low density lipoprotein by endothelial cells involves lipid peroxidation and degradation of low density lipoprotein phospholipids. *Proc. Natl. Acad. Sci. U.S.A.* **81**: 3883-3887.

Susi, H., 1969, Infrared Spectra of Biological Macromolecules and Related Systems. In *Structure and Stability of Biological Macromolecules.* (Timasheff,S.N.; and Stevens,L. eds.), Dekker. New York, pp. 575-663.

Sutherland, G.B.B.M., 1952, Infrared Analysis of the Structure of Amino Acids, Polypeptides and Proteins. In *Advances in Protein Chemistry* (Anson, M. L., Bailey, K., and Edsall, J. T. eds.). Academic Press. New York, pp. 291-318

Yang, C.Y., Chen, S.H., Gianturco, S.H., Bradley, W.A., Sparrow, J.T., Tanimura, M., Li, W. H., Sparrow, D.A., DeLoof, H., Rosseneu, M., Lee, F.S., Gu, Z.W., Gotto, A.M., Jr., and Chan, L., 1986, Sequence, structure, receptor-binding domains and internal repeats of human apolipoprotein B-100. *Nature* **323**: 738-742.

Yang, J.Y.N., Treleaven, W.D., and Cushley, R.J., 1991, An alternate apoprotein conformation in high density apolipoprotein discoidal complexes. A Fourier transform infrared study. *Biochem. Int.* **25**: 1077-1086.

Ylä-Herttuala, S., 1991, Macrophages and oxidized low density lipoproteins in the pathogenesis of atherosclerosis. *Ann. Med.* **23**: 561-567.

New Approaches in Spin Labeling and Spin Trapping. Part One: ESR Studies of Local Chemical Environment

VALERY V. KHRAMTSOV
Institute of Chemical Kinetics and Combustion, Novosibirsk 630090, Russia

1. INTRODUCTION

Spin labeling and spin trapping techniques are based on detection of ESR spectra of spin labels/probes or spin adducts which in many cases are nitroxyl radicals (NR). Spin traps have high reactivity towards free radical chemistry while spin labels are designed to prevent involvement of their paramagnetic fragment in chemical reactions. As a matter of fact the chemical reactions of nitroxides are often considered as a factor that compromises the stability of the paramagnetic fragment of NR and therefore limits their applications in spin labeling.

However NR were found to participate in series of the reactions of biological interest without loss of their paramagnetic properties (Khramtsov and Volodarsky, 1998). In this case the difference in ESR spectra of the radical-reagent and radical-product allows to measure the reactivity and concentration of the corresponding reactive substances. In particular ESR spectroscopy allows to measure the local concentration and reactivity of the proton (pH), thiol-or NO-donating compounds (Khramtsov and Volodarsky, 1998), which are of principal importance for many biological functions. The ESR spectra sensitivity of the NR to local chemical environment opens a new field of applications in spin labeling. Moreover the recent progress in low-field ESR-based imaging and spectroscopy-based techniques allows some of these specific spin probes to be used for functional *in vivo* measurement and imaging. For resent pertinent reviews, see Swartz and Berliner (1997), Mäder (1998) and Khramtsov *et al.* (2000).

Supramolecular Structure and Function 7, edited by Pifat-Mrzljak.
Kluwer Academic/Plenum Publishers, New York, 2001.

2. pH-SENSITIVE NITROXYL RADICALS

2.1 Basis of the method of spin pH probes

The determination of pH is one of the most important problems in the biochemistry of living organisms, since many of the vital processes of cells and cellular organelles depend on the local pH value. Amongst currently used experimental approaches for the measurement of pH, the method of spin pH probes is a comparatively new and rapidly developing field. A wide set of spin pH probes with different ranges of pH sensitivity, labeling group, and lipophilicity is currently available for application in various chemical and biological systems (Khramtsov and Volodarsky, 1998).

Stable NR of the imidazoline and imidazolidine types have been shown to be the most promising spin pH probes due to the presence of protonatable nitrogen atom N-3 in the radical heterocycle and therefore due to the large effect of pH on their ESR spectra (Keana *et al.*, 1982; Khramtsov *et al.*, 1982). Figure 1a demonstrates the reversible influence of pH on X-band ESR spectra of the imidazolidine radical R1 (Scheme 1). The ESR spectrum at pH close to pK of the radical arises from the superposition of two triplet spectra with different hyperfine interaction constants, $a_N(R)$ and $a_N(RH+)$, and g-factors (Δa_N=1.25 G and Δg=0.0003) for R and RH+ forms which participate in proton exchange according to reaction shown in Scheme 1:

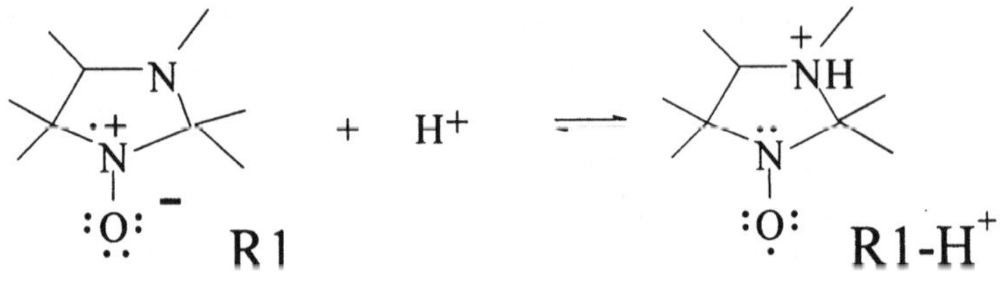

Scheme 1. Reaction of proton exchange between protonated and unprotonated forms of the radical R1. The main resonance structures of nitroxide moiety are shown indicating the increase of unpaired electron density at the nitrogen atom N-1, and therefore, higher value of a_N constant for the unprotonated form.

The pH dependence of the relative contribution of protonated, RH+, and unprotonated, R, forms in the ESR spectrum (see Figure 1b) provides the basis of the application of this nitroxide for pH determination. Fitting the experimental data to a calculated curve yields a good agreement using conventional titration equation:

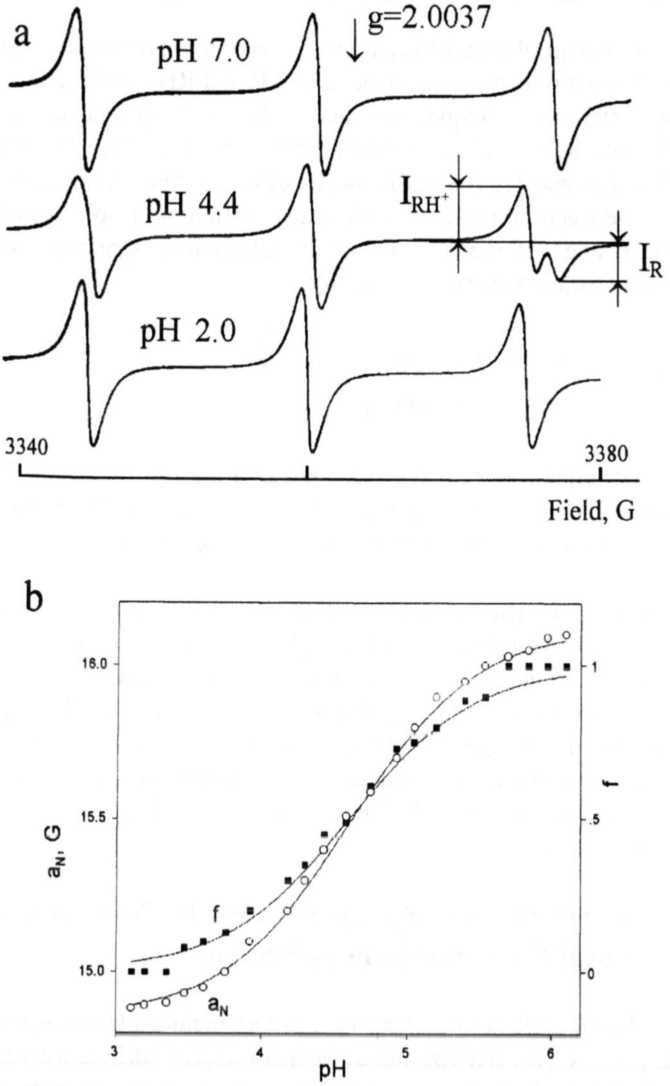

Figure 1. (a): The X-band ESR spectra of aqueous solution of imidazolidine radical R1 at different pHs. *(b):* The pH dependence of the spectral intensity ratio, $f=I_R/(I_R+I_{RH+})$ and hyperfine interaction splitting, a_N, measured as the distance between the baseline crossing points of the low-field and central lines of the spectrum. Solid curves were calculated using equation (1) assuming $I_R/I_{RH+} = [R]/[RH+]$, and (2) with the parameters of pK = 4.6, $a_N(R) =$ 16.12 G and $a_N(RH^+) = 14.85$ G.

$$[RH^+]/[R] = [H^+]/K_a \qquad (1)$$

where K_a is an equilibrium constant of the reaction of proton exchange (pK= - logK_a). The more distinguished signals of R and RH+ forms are observed at higher microwave frequencies due to frequency-dependent Δg effect while this resolution is less at lower ESR frequencies (see Fig.2 for 140 GHz, 35.5 GHz and 1.2 GHz ESR spectra of R1). In the latter case of proton exchange between unresolved ESR lines, variation of pH results in a reversible change of the a_N constant in satisfactory agreement with the conventional titration curve (see Fig.1b):

$$a_N(H^+) = \frac{a_N(R) + a_N(RH^+) \times [H^+]/K_a}{1 + [H^+]/K_a} \qquad (2)$$

Note, that equation (2) is precisely valid for the fast frequency R⇔RH+ exchange observed in ESR spectra of various pH-sensitive nitroxides in acidic (pH<3) or alkaline (pH>11) aqueous solutions (Khramtsov *et al.*, 1985).

Availability of a wide variety of pH-sensitive NR with different ranges of pH-sensitivity, labeling group and lipophilicity facilitates their application to a variety of biological systems from subcellular organelles to complex organisms. Moreover, isotopic substitution, especially introducing of ^{15}N and/or ^{2}H into the structure of pH-sensitive radical (Glazachev *et al.*, 2000) is a promising approach to enhancing the information obtained from spin labeling experiments due to the better resolution and intensity of the ESR spectra (see Fig.3).

2.2 Applications of Spin pH Probes for Measurement of Local Concentrations of Protons

X-band ESR applications of spin pH probes include pH-measurements in aqueous phase of the heterogeneous microstructures such as micropores of cross-linked polyelectrolites (Molochnikov et al., 1996) or mesopores of zeolites and kaolin (Zamaraev *et al.*, 1997). Biological ESR applications at X-band are mostly limited by *in vitro* or *ex vivo* studies due to high nonresonant absorption of the excitation frequency by aqueous samples. Therefore spin pH probes have been used for the studies of the samples whose volume does not exceed 200 μl such as liposome solutions, water-in-oil (w/o) ointments, proteins and phospholipid membranes. Specific nitroxide structures have been designed for these studies.

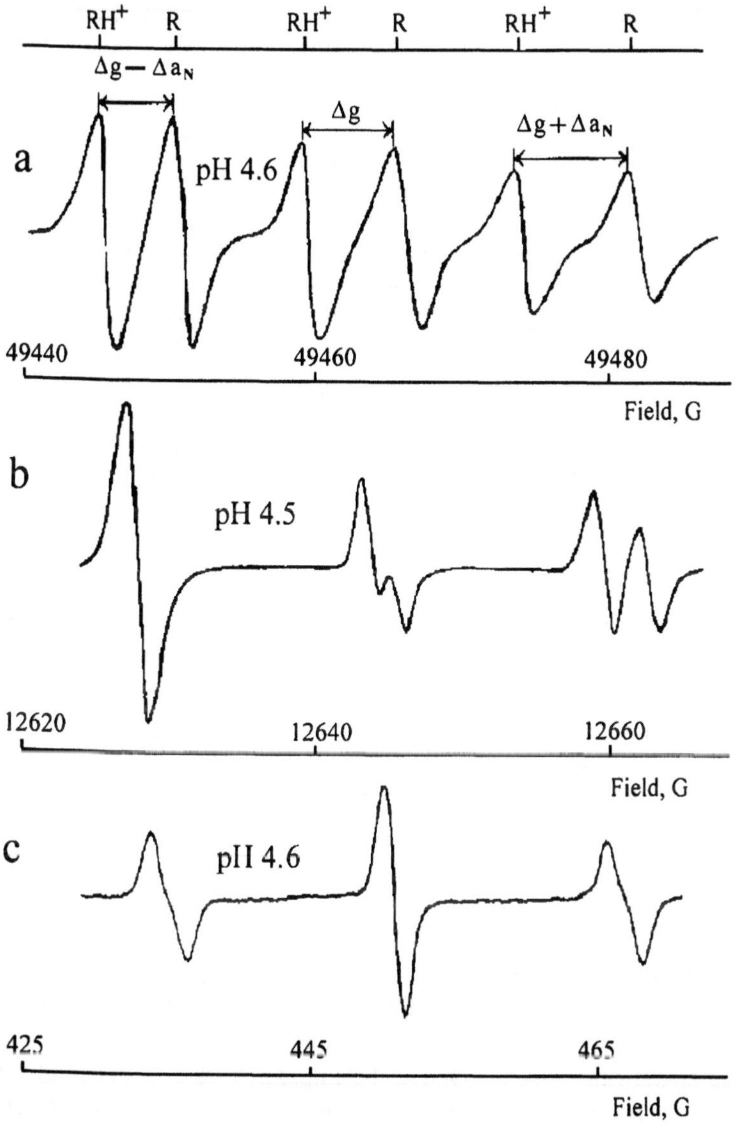

Figure 2. The ESR spectra of an aqueous solution of the nitroxide R1 at a pH close to the pK of the radical at various microwave frequencies: 140 GHz (a), 35.5 GHz (b) and 1.2 GHz (c). The scheme at the top illustrates the origin of the ESR line positions from superposition of two triplet spectra of protonated and unprotonated forms of the radical with different values of g-factor and a_N constant.

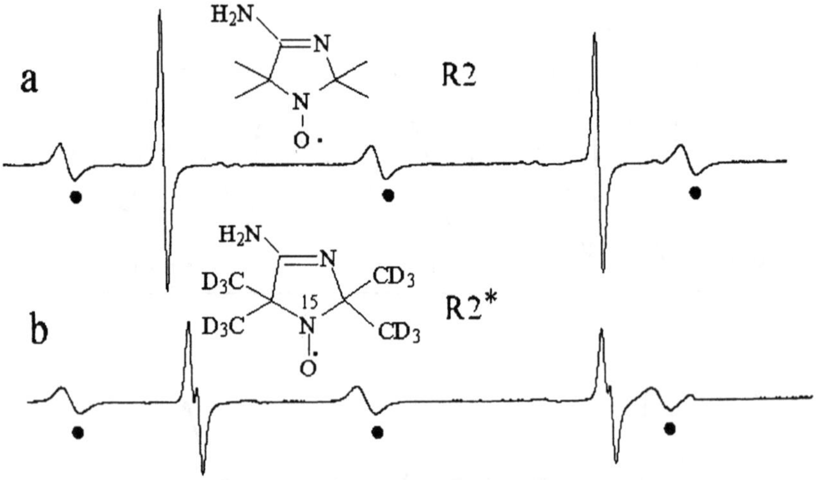

Figure 3. The X-band ESR spectra of nitrogen bubbled aqueous solutions of mixture of the radicals R2 and R2* at equal concentrations of 0.3 mM at 23°C and at pH 8 (a), and 4 (b). The symbols (•) denoted ESR spectral lines of the radical R2. Note that isotopic substitution results in increase in peak intensities of ESR spectrum of R2* compared with that for R2 in about 7 times for unprotonated form (a) and in about 5 times for protonated form (b). The narrowing of peak-to-peak linewidths, ΔH_{p-p}, is from 0.76 G to 0.33 G for R and from 0.97 G to 0.7 G for RH+ forms, respectively. Moreover decrease in ΔH_{p-p} for R2* results in uncovering of an additional doublet splitting of 0.36 G with attached proton (b). The ESR spectra setting were as following: microwave power 5 mW, modulation amplitude 0.3 G, receiver gain 10^4.

Kroll et al. (1995) applied the highly hydrophilic imidazoline spin pH probe R2 (see Fig.3 for the structure) for pH monitoring in the water phase of w/o systems during decomposition of hydrolyzable drugs. Real-time pH monitoring of this drug delivery system using glass electrodes or fluorescent molecular indicators were found to be impossible due to nonconductivity and nontransparency of the samples while applications of spin pH probes allows to overcome these limitations.

Determining intraliposomal pH is of basic interest due to the wide spectrum of liposome applications, including drug delivery and studies of biomembrane transport processes. For localization of spin pH probes into the inner aqueous volume of liposomes and preventing their efflux through the hydrophobic phospholipid moiety, polyanionic substitutes were incorporated into the structures of the radicals R1 and R2 (Khramtsov *et al.*, 1989; Balakirev *et al.*, 1992, Balakirev and Khramtsov, 1993). Incorporation of a long hydrocarbon tail into the structure of R1 has been used to synthesize the hydrophobic spin pH probe for localization in the lipid phase of biomembranes and measuring the surface potential and polarity of phospholipid membranes (Khramtsov *et al.*, 1992). Structural analogs of radicals R1 and R2 with alkylating, carboxy and disulfide groups were used for protein labeling (Khramtsov *et al.*, 1990, 1992, 1997) resulting in the formation of macromolecular pH-sensitive probes.

The application of spin pH probes offers a unique opportunity for non-invasive pH assessment in living animals using low-field ESR-based techniques (for reviews see Mäder *et al.*, 1998, Khramtsov *et al.*, 2000). The L-band ESR spectroscopy (1.2 GHz) was shown to be a valuable tool for *in vivo* monitoring of biodegradable drug delivery systems (Mäder *et al.*, 1996) and pH measurement in the mouse gut (Gallez *et al.*, 1996). It is necessary to work at even lower ESR frequency when whole body studies of larger animals such as the rat are required. pH effect on the radical Rl in solution has been demonstrated by Sotgiu *et al.* (1996, 1998) using spectral-spatial and spatial-spatial methods at ESR frequencies down to 280 MHz. Recently spin pH probes were used for the studies in living rats (Foster *et al.*, 1999, Khramtsov *et al.*, 2000) using Longitudinally Detected ESR (LODESR) (Nicholson *et al.*, 1994) and the double resonance technique of Field-Cycled Dynamic Nuclear Polarisation (FC-DNP) (Lurie et al., 1991) with its imaging analogue Field-Cycled Proton-Electron Double-Resonance Imaging (FC-PEDRI) (Foster *et al.*, 1998, Lurie, 2000) operating at an ESR frequency as low as 120 MHz.

3. ESR DETECTION OF THIOL STATUS

3.1 Basis of the method of thiols determination using paramagnetic analogs of Ellman's reagent

The key role of thiols and thiol-disulfide status of the cell in a wide array of biochemical and biological responses including that in response to oxidative stress has been amply documented (Sies, 1999). Optical and chromatographic methods have been used for quantitative determination of thiols (Boyne and Ellman, 1972; Kosower and Kosower, 1995; Newton and Fahey, 1995). The requirement of optical transparency of the samples and labor consuming procedures in case of chromatographic methods, in particular HPLC, are the main limitations of these methods. We proposed ESR methods for quantitative determination of sulfhydryl groups (Khramtsov *et al.*, 1989, 1997) which allows overcome some of these limitations. The method is based on the application of disulfide biradicals, RSSR, which participate in thiol-disulfide exchange reactions with thiols, BSH:

$$RSSR + BSH \Leftrightarrow RSH + BSSR \qquad (3)$$

The reaction (3) follows by distinct alterations in the ESR spectrum, since a three-line monoradical spectrum results from the five-line biradical starting material (see Fig.4). The linear dependence of the relative changes of the peak intensities of monoradical ($\Delta I_m/I_m^0$) and biradical ($\Delta I_b/I_b^0$) components of ESR spectra of these disulfide probes on the concentration of thiols was observed at low thiol concentrations (Khramtsov *et al.*, 1989, 1997). This linear part of the dependencies is described by the equations:

$$\Delta I_b / I_b^0 = -[BSH]/[RSSR]; \quad \Delta I_m / I_m^0 = \alpha[BSH]/[RSSR] \qquad (4)$$

The tangent of the angle of the slope of the dependence of $\Delta I_b/I_b^0$ on [BSH]/[RSSR] is equal to one in agreement with the stoichiometry of the reaction (3) and does not depend on the nature of thiol, B-SH, or biradical, RSSR. The slope, α, of the dependence ($\Delta I_m/I_m^0$) depends on the integral intensity of the biradical ESR signal compared with that of monoradicals and, as consequence, on the nature of both the thiol and biradical. In the case of GSH and cystein similar thiol-induced ESR changes were observed supporting the weak influence of the nature of low-molecular weight thiols on the ESR spectra of the resulted monoradical BSSR. The applications of the ratio $\Delta I_m/I_m^0$ for thiols quantification is more favorable due to higher sensitivity (usually $\alpha >> 1$, e.g. see Fig.4 for GSH) while the advantage of the ratio $\Delta I_b/I_b^0$ is independence on thiol structure (see Khramtsov and Volodarsky, 1998, for discussion).

3.2 Applications of the biradical disulfides for measurement of local concentrations of thiols

The biradical R_1SSR_1 has been used to measure GSH content in erythrocytes (Khramtsov *et al.*, 1989, 1997), Chinese hamster ovary cells (Weiner *et al.* 1991), murine neuroblastoma, malignant melanoma cells (Busse *et al.*, 1992a), and tumor HeLa cells (Busse *et al.*, 1992b) during the treatment by anticancer antibiotics. The ESR method for intracellular GSH determination is based on the dominant contribution of GSH in the reaction of the biradical with thiols, fast diffusion of the biradical across the lipid membranes and its comparatively low reduction rate. The sensitivity of the method allows GSH measurements in a few (~100) cells (Weiner *et al.* 1991). ESR measurements of GSH in humans and rats under oxidative stress conditions showed increased levels of oxidized glutathione in the blood plasma, e.g. in some human pathologies, such as kyphoscoliosis (Yelinova *et al.*, 1996). The ESR measurements of the thiols in the blood using biradical disulfide appears to

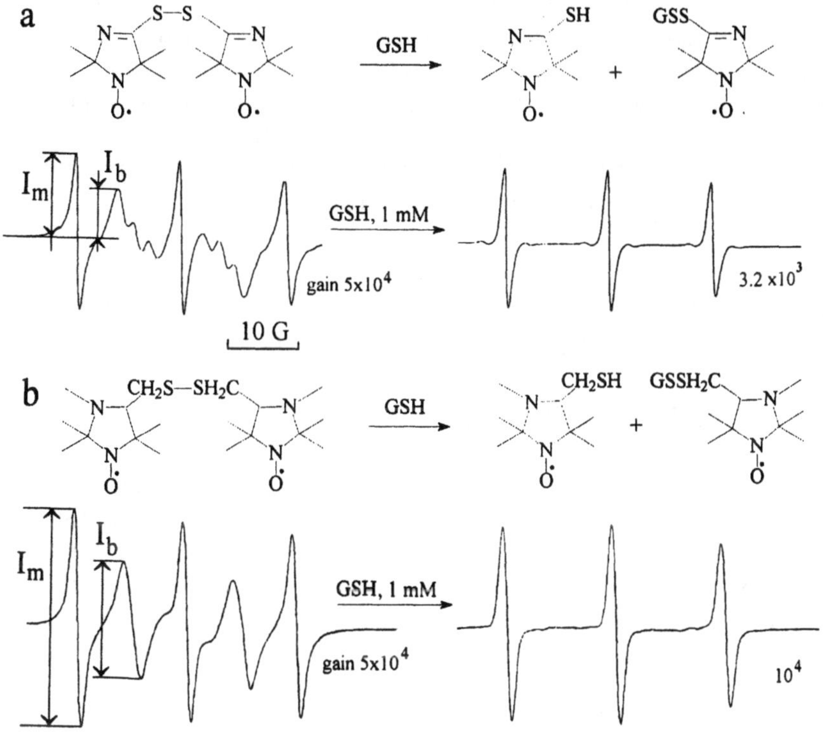

Figure 4. The chemical structures and ESR spectra of 0.1 mM solutions of the two disulfide biradicals, R_1SSR_1 in 10 mM Na-phospate buffer, pH 7.5 (a), and R_2SSR_2 in 10 mM $Na_2B_4O_7$ buffer, pH 11.3 (b). Spectra transformations upon the addition of excess of GSH (1 mM) are shown. A double integration of the ESR spectra of the biradicals supports preservation of their integral intensity upon the reaction with GSH and, therefore, indicates insignificance of the nitroxide moiety reduction by SH groups. Spectrometer settings were the same except for the gain indicated near the spectra: microwave power 20 mW, modulation amplitude 0.8 G.

be a sensitive, convenient method which does not need complicated sample preparation.

The application of the biradical R_1SSR_1 for the determination of thiol levels in isolated perfused hearts undergoing oxidative stress induced by ischemia/reperfusion was described by Nohl *et al.* (1995). The authors measured release of the monoradicals into the perfusate due to the reaction of the biradical disulfide with tissue thiols and observed the decrease in the reduced GSH level following ischemia/reperfusion. One of the disadvantages of this approach is the requirement of the excess of the label over thiols resulting in the "consumption" of critical SH groups followed by irreversible damage of the system. To overcome this limitation we proposed recently a new disulfide biradical R_2SSR_2 (Khramtsov *et al.*, 1997). The applications of this biradical disulfide allows much lower label concentrations and therefore does not result in significant consumption of thiols. Recently Yelinova *et al.* (1999) applied low noninvasive concentrations of R_2SSR_2 for demonstration of increased GSH oxidation in animal model of stress-sensitive arterial hypertension.

4. EPR DETECTION OF NITRIC OXIDE USING NITRONYLNITROXYL RADICALS (NNR)

4.1 Basis of NNR applications as NO traps

The ESR methods of NO detection in solution are based on the trapping of nitric oxide with formation of comparatively stable paramagnetic species using oxyhemoglobin, $HbFe^{II}$, with formation of the nitrosyl derivative $HbFe^{II}$-NO (Naschimento *et al.*, 1991); metal-chelator complexes of reduced iron (Fe^{2+}) with dithiocarbamate derivatives (Mordvintcev *et al.*, 1991); cheletropic spin traps (Korth *et al.*, 1994) and nitronylnitroxyl radicals which react with NO forming stable iminonitroxides (Akaike *et al.*, 1993; Joseph *et al.*, 1993; Woldman *et al.*, 1994).

Nadeau and Boocock (1977) has described NNR applications for ESR measurement of NO in gas phase. The applications of different NNR for NO detection in aqueous solutions has been proposed by Akaike *et al.* (1993), Joseph *et al.* (1993) and Woldman *et al.* (1994). Fig.5 demonstrates the typical changes of the ESR spectrum of one of the NNR induced by the addition of NO-saturated aqueous solution. The ESR spectra of the NNR consist of a characteristic quintet of lines with an intensity ratio 1:2:3:2:1 due to the similar hyperfine interaction constants with nuclear spins of atoms N-1 and N-3 (a_{N1}=a_{N3}=8.1 G, see Fig.5). Addition of NO-saturated buffer to NNR solution results in the formation of distinctly different ESR spectrum of seven lines (a_{N1}=9.8 G, a_{N3}=4.6 G). The latter spectrum corresponds to imino nitroxyl radical, INR. This spectral transformation proceeds with preservation of the integral intensity supporting stoichiometric conversion of NNR into the corresponding INR by the reaction shown in Fig.5.

Note that transformation of NNR into INR is specific reaction of nitronyl nitroxides with nitric oxide and was not observed in the presence of other radical species as $^{\bullet}NO_2$, $O_2^{\bullet -}$ or $^{\bullet}OH$ (Akaike *et al.*, 1993; Joseph *et al.*, 1993).

The rate constants for the reaction of the NNR with nitric oxide are about 10^4 $M^{-1}s^{-1}$ (Akaike *et al.*, 1993, Woldman *et al.*, 1994). Therefore the characteristic time of NO trapping by NNR ($\tau \approx 1/k_1[NNR] \leq 1$ s at $[NNR] \geq 0.1$

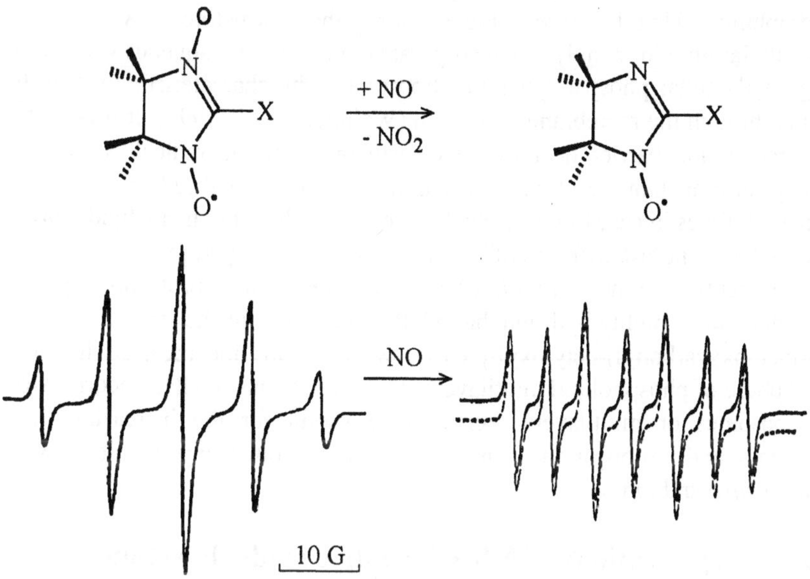

Figure 5. Scheme of the NO detecting reaction. NNR reacts with NO to produce INR. The corresponding typical changes in ESR spectra from a five-line pattern to a seven-line pattern are shown for 0.15 mM solution of one of the NNR (NNR1, X = 4-trimethylammoniophenyl) in 50 mM K-phosphate buffer, pH 7.4, after adding NO-saturated solution with a final NO concentration 0.35 mM. Dotted line: the ESR spectrum of 0.15 mM of corresponding INR. Since the low-field components of the ESR spectra of NNR and INR did not overlap, the kientics of the reaction can be followed by time evolution of one of these components.

mM) is close to or less than lifetime of NO *in vivo* (Ignarro, 1990). The ability to follow transformation of NNR into INR by ESR provides the basis for quantitative measurements of concentration of trapped nitric oxide, the sensitivity of the approach is 1 μM for detection of NO concentration and 0.3 nM/s for the measurements of the rates of NO generation for 1h in the sample of 200 μl. It has an advantage of one order higher sensitivity compared with widely used spin trapping of NO by dithiocarbamate complexes of iron (Mordvintcev *et al.*, 1991; Komarov *et al.*, 1993) due to comparatively narrow lines of ESR spectra of NNR and INR.

The applications of NNR for NO detection by ESR in biological systems is limited due to the fast reduction of NNR into diamagnetic product. The reduction rates of NNR were found to be much higher compared with that for di-tert-alkylnitroxides (Woldman, 1994; Haseloff, 1997). To overcome this limitation we proposed to use NNR incorporated in phospholipid liposomes (Woldman *et al.*, 1994) or multilayers (Yelinova *et al.*, 2000). The highly hydrophilic NNR1 (see Fig.5 for the structure, X = 4-trimethylammoniophenyl) was incorporated into the inner aqueous volume of large unilamellar phospatidylcholine liposomes with characteristic time of its efflux through the membrane, $\tau_{1/2} \approx 10$ h (Woldman *et al.*, 1994). It was found that membrane barrier decreases the availability of the radical NNR1 for biological reductants resulting in strong decrease of radical reduction rates in about 10^3 times in case of rat cerebella cytosol. On the other hand, lipid bilayer does not prevent fast diffusion of low-molecular-weight hydrophobic molecule, NO. Recently Yelinova *et al.* (2000) observed strong protection against bioreduction of highly hydrophobic NNR2 (see Fig.5 for the structure, X= 4-[(8-methoxycarbonyl)-octyloxy]-phenyl) when it was incorporated into the lipid phase of phospholipid multilayers. The applications of NNR incorporated in liposomes or multilayers seems to be useful for the increasing their efficiency and prolongation of their action as antagonists against nitric oxide both *in vitro* and *in vivo*.

4.2 Applications of NNR for nitric oxide detection

NNR have been used to study nitric oxide release from NO donors both in chemical and biological systems. Joseph *et al.* (1993) have applied NNR to study NO release during photolytic decomposition of sodium nitroprusside. NNR were successfully used to study the mechanisms of NO release from S-nitroso and C-nitroso compounds during their photodecomposition (Singh *et al.*, 1995) as well as thiol-induced NO release from S-nitroso adducts of glutathione and bovine serum albumin (Pietraforte *et al.*, 1995). We have used NNR traps to study NO liberation by 3,4-dihydro-1,2-diazete 1,2-dioxides (Utepbergenov *et al.*, 1995; Khramtsov *et al.*, 1996; Kiriljuk *et al.*, 1998), by NN-bis(arylsulfonyl)hydroxylamines (Balakirev and Khramtsov, 1996) and by the derivatives of nitrosometoxyamines (Khlestkin *et al.*, 1996). Fig.6 demonstrates the typical transformation of the low-field component of the ESR spectrum of NNR1 during the incubation with the NO donor, which allows one to calculate the rate constant of the process.

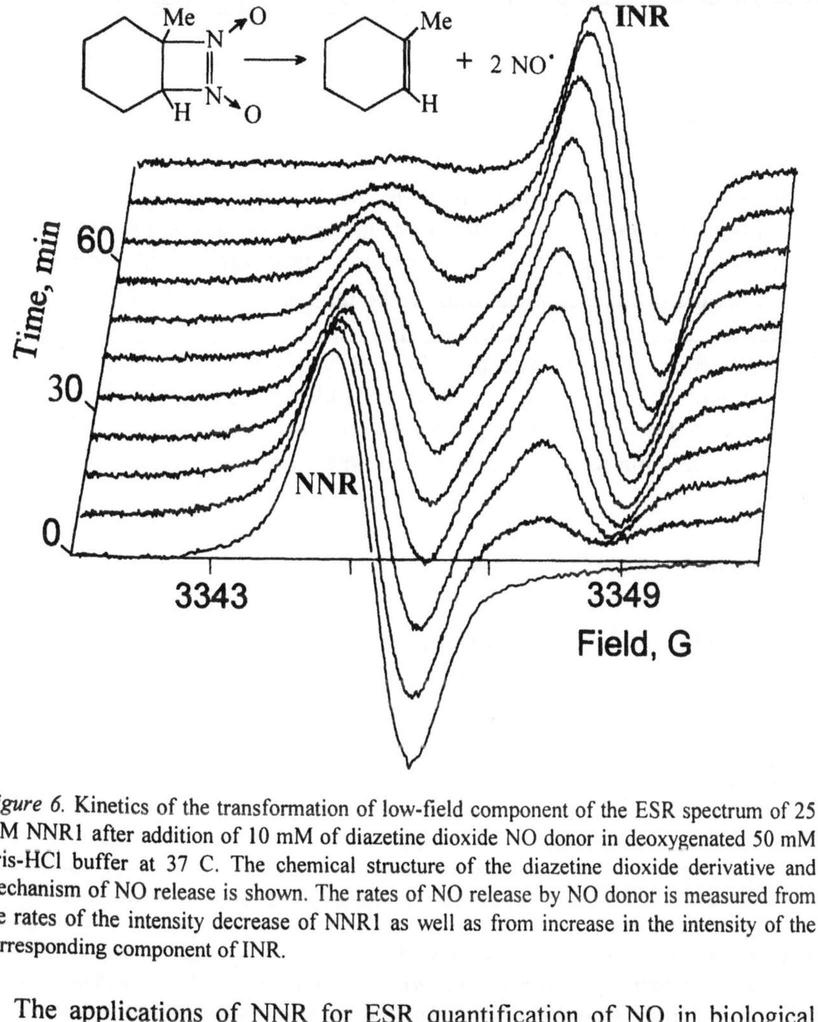

Figure 6. Kinetics of the transformation of low-field component of the ESR spectrum of 25 μM NNR1 after addition of 10 mM of diazetine dioxide NO donor in deoxygenated 50 mM Tris-HCl buffer at 37 C. The chemical structure of the diazetine dioxide derivative and mechanism of NO release is shown. The rates of NO release by NO donor is measured from the rates of the intensity decrease of NNR1 as well as from increase in the intensity of the corresponding component of INR.

The applications of NNR for ESR quantification of NO in biological system is limited by their fast reduction. Nevertheless Woldman *et al.* (1994) has demonstrated applicability of NNR for measurement of arginine dependent production of nitric oxide in tissue homogenates in rat cerebella cytosol *in vitro*. To decrease bioreduction of the radical fragment the NNR1 incorporated into the liposomes was applied. Yoshida *et al.* (1994) observed NO-dependent transformation of the NNR into INR in lipopolysaccharide-treated rats when studying endotoxin shock *in vivo*. The main fraction (about 80%) of the NNR injected in the rats was observed in the urine in diamagnetic forms (mostly hydroxylamines) which were reoxidised to the corresponding NNR and INR with 1 mM potassium ferricyanide for ESR detection.

5. CONCLUSION

It is apparent that stable nitroxides with ESR spectra sensitivity to concentration and reactivity of different chemical species open a new field of applications of the spin labeling method. In this respect imidazoline and imidazolidine nitroxides were found to be the most promising providing the possibility to follow their reactions with protons, thiols and nitric oxide by ESR. The future development of stable NR for the studies of local chemical environment has a potential value to extend both the ESR spin labeling applications described in this review as well as a number of the chemical functions available for evaluation by ESR spectroscopy.

ACKNOWLEDGMENTS

The author wish to acknowledge funding for this research in part from the grants of INTAS (Project 1999-1086) and NWO (047.006.010).

REFERENCES

Akaike, T., Yoshida, M., Miyamoto, Y., Sato, K., Kohno, M., Sasamoto, K., Miyazaki, K., Ueda, S., and Maeda, H., 1993, Antagonistic action of imidazolineoxyl N-oxides against endothelium-derived relaxing factor/ NO through a radical reaction. *Biochemistry* **32**: 827-832.

Balakirev, M., and Khramtsov, V., 1993, New pH-sensitive aminoxyls - application to the study of biomembrane transport processes, *J.Chem.Soc. Perkin Trans.* **2**/11: 2157-2160.

Balakirev, M.Yu, and Khramtsov, V.V., 1996, EPR study of free radical decomposition of NN-bis(arilsulfonil)hydroxylamines in organic solution. *J.Org.Chem.* **61**: 7263-7269.

Balakirev, M.Yu., Khramtsov, V.V., Berezina, T.A., Martin, V.V., and Volodarsky L.B., 1992, The synthesis of amidine derivatives of imidazoline nitroxides - a new series of pH-sensitive spin probes and labels, *Synthesis* **12**: 1223-1225.

Boyne, A. F., and Ellman, G. L. ,1972, A methodology for analysis of tissue sulfhydryl components. *Anal. Biochem.* **46:** 639-653.

Busse, E., Zimmer, G., and Kornhuber, B., 1992a, Plasma-membrane fluidity studies of murine neuroblastoma and malignant melanoma cells under irradiation. *Strahlenther Onkol.* **168:**419-422.

Busse, E., Zimmer, G., Schopohl, B., and Kornhuber, B., 1992b, Influence of alpha-lipoic acid on intracellular glutathione in vitro and in vivo. *Arzneim.-Forsch./Drug Res.* **42:**829-831.

Foster, M.A., Grigor'ev, I.A., Lurie, D.J., Khramtsov, V.V., Hutchison, J.M.S., McCallum, S., Panagiotelis, I., and Nicholson, I., 1999, Further low-field ESR studies with the pH-sensitive spin probe HMI. In *Abstr. Intern. Workshop Biomed. Applications of In Vivo EPR and PEDRI*, Aberdeen (Scotland), pp. 49.

Foster, M.A., Seimenis, I., and Lurie, D.J., 1998, The application of PEDRI to the study of free radicals *in vivo*. *Phys.Med.Biol.* **43**: 1893-1897.

Gallez, B., Mäder, K., and Swartz, H.M., 1996, Noninvasive measurement of the pH inside the gut by using pH-sensitive nitroxides: An *in vivo* EPR study. *Magn. Reson. Med.* **36**: 694-697.

Glazachev, Yu.I., Grigor'ev, I.A., Foster, M.A., Reijerse, E.J., and Khramtsov, V.V., 2000, EPR studies of ^{15}N and ^{2}H-substituted imidazoline spin pH probe. In *Abstr. of Seventh Intern. Summer School on Biophysics*, Rovinj (Croatia), 2000, pp.85.

Haseloff, R.F., Zöllner, S., Kiriljuk, I.A., Grigor'ev, I.A., Reszka, R., Bernhardt, R., Mertsch, K., Roloff, B., and Blasig, I.E., 1997, Superoxide-mediated reduction of the nitroxide group can prevent detection of nitric oxide by nitronyl nitroxides. *Free Radical Res.* **26**: 7-17.

Ignarro, L.J., 1990, Biosynthesis and metabolism of endothelium-derived nitric oxide. *Annu.Rev.Pharmacol. Toxicol.* **30**: 535-560.

Joseph, J., Kalynaraman, B., and Hyde, J.S., 1993, Trapping of nitric oxide by nitronyl nitroxides: an electron spin resonance investigation. *Biochem.Biophys.Res.Commun.* **192**: 926-934.

Keana, J.F.W., Acarregui, M.J. and Boyle, S.L.M., 1982, 2,2-Disubstituted-4,4-dimethylimidazolidinyl-3-oxy nitroxides indicators of aqueous acidity through variation of a_N with pH. *J.Am.Chem.Soc.* **104**: 827-830.

Khlestkin, V.K, Mazhukin, D.G, Tikhonov, A.Ya, Bagryanskaja, I.Yu., Gatilov, Yu.V, Utepbergenov, D.I, Khramtsov, V.V., and Volodarsky, L.B., 1996, Unexpected transformation of 1,2-bis(N-methoxy-N-nitrosoamino)cycloalkanes: First synthesis of 4,5-dihydro-1,2,3-triazole 2-oxides. *Tetrahedron Lett.* **37**, 5997-6000.

Khramtsov, V.V., and Volodarsky, L.B., 1998, Use of imidazoline nitroxides in studies of chemical reactions: ESR measurements of concentration and reactivity of protons, thiols and nitric oxide. In *Biological Magnetic Resonance, Volume 14: Spin Labeling: The Next Millennium* (L.J.Berliner, ed.), Plenum Press, New York. pp.109-180.

Khramtsov, V.V., Grigor'ev, I.A., Foster, M.A., Lurie, D.J., and Nicholson, I., 2000, Biological applications of spin pH probes. *Cell.Mol.Biol.* **46/8**: 1361-1374.

Khramtsov, V.V., Marsh, D., Weiner, L.M., and Reznikov, V.A., 1992, The application of pH-sensitive spin labels to studies of surface potential and polarity of phospholipid membranes and proteins. *Bioch.Bioph.Acta,* **1104**: 317-324.

Khramtsov, V.V., Panteleev, M.V., and Weiner, L.M., 1989, ESR study of proton transport across phospholipid vesicle membranes. *J.Bioch.Biophys.Methods*, **18**: 237-246.

Khramtsov, V.V., Utepbergenov, D.I., Woldman, Ya.Yu., Vlassenko, L.P., Markel, A.L., Mazhukin, D.G., Kiriljuk, I.A., Grigor'ev, I.A., Tikhonov, A.Ya., and Volodarsky, L.B., 1996, In vitro and in vivo studies of the derivatives of 1,2-diazetine and nitronylnitroxides as donors and acceptors of nitric oxide. *Biochemistry (Moscow)* **61**: 1223-1231.

Khramtsov, V.V., Weiner, L.M., Eremenko, S.I., Belchenko, O.I., Schastnev, P.V., Grigor'ev, I.A., and Reznikov, V.A., 1985, Proton exchange in stable nitroxyl radicals of imidazoline and imidazolidine series. *J. Magn. Res.*, **61**: 397-408.

Khramtsov, V.V., Weiner, L.M., Grigor'ev, I.A., and Volodarsky, L.B., 1982, Proton exchange in stable nitroxyl radicals. EPR study of the pH of aqueous solutions. *Chem.Phys.Lett.* **91**: 69-72.

Khramtsov, V.V., Weiner, L.M., Rar, V.A., Berezina, T.A., Martin, V.V., and Volodarsky, L.B., 1990, Macromolecular spin pH probes based on human serum albumin. *Biochemistry (Moscow)* **55**: 1014-1017.

Khramtsov, V.V., Yelinova, V.I., Glazachev, Yu.I., Reznikov, V.A., and Zimmer, G., 1997, Quantitative determination and reversible modification of thiols using imidazolidine biradical disulfide label. *J. Biochem.Biophys. Methods*, **35/2**: 115-128.

Khramtsov, V.V., Yelinova, V.I., Weiner, L.M., Berezina, T.A., Martin, V.V., and Volodarsky, L.B., 1989, Quantitative determination of SH groups in low- and high-molecular-weight compounds by an electron spin resonance method. *Anal. Biochem.* **182**:58-63.

Kirilyuk, I.A., Utepbergenov, D.I., Mazhukin, D.G., Fechner, K., Mertsch, K., Khramtsov, V.V., Blasig, I.E., and Haseloff, R.F., 1998, Thiol-Induced Nitric Oxide Release from 3-Halogeno-3,4-dihydrodiazete 1,2-Dioxides. *J.Med.Chem.*, **41/7**: 1027-1033.

Komarov, A., Mattson, D., Jones, M.M., Singh, P.K., and Lai, C.S., 1993, In vivo spin trapping of nitric oxide in mice.*Biochem.Biophys.Res.Commun.* **195**: 1191-1198.

Korth, H.-G., Sustmann, R., Lommes, P., Paul, T., Ernst, A., de Groot, H., Hughs, L., and Ingold, K.U., 1994, Nitric oxide cheletropic traps (NOCTs) with improved thermal stability and water solubility. *J.Am.Chem.Soc.* **116**: 2767-2770.

Kosower, N.S., and Kosower, E.M., 1995, Diamide: an oxidant probe for thiols. *Methods Enzymol.* **251**:123-133.

Kroll, C., Mäder, K., Stößer, R. and Borchert, H.H., 1995, Direct and continuous determination of pH values in nontransparent w/o systems by means of EPR spectroscopy. *European J. Pharmaceutical Sciences* **3**: 21-26.

Lurie, D.J., Nicholson, I., and Mallard, J.R., 1991, Low-field EPR measurements by field-cycled dynamic nuclear polarisation. *J.Magn.Reson.* **95**: 405-409.

Lurie, D.J., 2000, Proton-electron double-resonance imaging (PEDRI). In *Biological Magnetic Resonance* (L.J. Berliner, ed.), vol. 18, Plenum Press, New York.

Mäder, K., 1998, Pharmaceutical applications of *in vivo* EPR. *Phys.Med.Biol.* **43**: 1931-1935.

Mäder, K., Gallez, B., Liu, K.J. and Swartz, H.M., 1996, Non-invasive *in vivo* characterization of release processes in biodegradable polymers by low-frequency electron paramagnetic resonance spectroscopy. *Biomaterials* **17**: 457-461.

Molochnikov, L.S., Kovalyova, E.G., Grigor'ev, I.A., and Reznikov, V.A., 1996, Determination of acidity in the interior of the cross-linked polyelectrolyte grain by the use of pH-sensitive probes. In *Metal-Containing Polymeric Materials* (C.U.Pittman, C.E.Carraher, M.Zeldin, J.E.Sheats and B.M.Culbertson, eds.), Plenum Press, New York and London, pp.395-401.

Mordvintcev, P., Mulsh, A., Busse, R., and Vanin, A., 1991, On line detection of nitric oxide formation in liquid aqueous phase by Electron Paramagnetic Resonance spectroscopy. *Anal.Biochem.* **199**: 142-146.

Nadeau, J.S., and Boocock, D.G.B., 1977, Stable free radical reagent and solid phase suitable for a nitric oxide dosimeter. *Anal.Chem.*, **49**: 1672-1676.

Naschimento, L.M., Neto, L.M., and Wajnberg, J., 1991, Tunneling within localized states in nitrosyl myoglobin. *J.Chem.Phys.* **95**: 2265-2268.

Newton, G.L., Fahey, R.C., 1995, Determination of biothiols by bromobiamane labeling and high-performance liquid chromatography. *Methods Enzymol.* **251**:148-166.

Nicholson, I., Robb, F.J.L., and Lurie, D.J. 1994, Imaging paramagnetic species using radiofrequency longitudinally detected ESR (LODESR imaging), *J.Magn.Reson. Series B*, **104**: 284 -288.

Nohl, H., Stolze, K., and Weiner, L.M. 1995, Noninvasive measurement of thiol status levels in cells and isolated organs. *Methods Enzymol.* **251**:191-203.

Pietraforte, D., Mallozzi, C., Scorza, G. and Minetti, M., 1995, Role of thiols in the targeting of S-nitroso thiols to red blood cells. *Biochemistry* **34**: 7177-7185.

Sies, H., 1999, Glutathione and its role in cellular function. *Free Radic. Biol. Med.* **27**: 916-921.

Singh, R.J., Hogg, N., Joseph, J., and Kalyanaraman, B., 1995, Photosensitized decomposition of S-nitrosothiols and 2-methyl-2-nitrosopropane. Possible use for site-directed nitric oxide production. *FEBS Lett.* **360**: 47-51.

Sotgiu, A., Mäder, K., Alecci, M., Ursini, C.L. and Placidi, G., 1996, EPR spectroscopy/imaging at RF (280 MHz) with pH-sensitive paramagnetic probes: a model study for *in vivo* applications, *Proc.Internat.Soc.Magn.Reson.Med.*, **3**: 1375.

Sotgiu, A., Mäder, K., Placidi, G., Colacicchi, S., Ursini, C.L. and Alecci, M., 1998, pH-sensitive imaging by low-frequency EPR: a model study for biological applications. *Phys.Med.Biol.* **43**: 1921-1930.

Swartz, H.M., and Berliner, L.J., 1997, In vivo EPR. In: *Foundations of Modern EPR* (G.R.Eaton and S.Eaton, eds.), World Scientific Publ. pp.362-380.

Utepbergenov, D.I., Khramtsov, V.V., Vlassenko, L.P., Markel, A.L., Mazhukin, D.G., Tikhonov, A.Ya., and Volodarsky, L.B. 1995, Kinetics of nitric oxide liberation by 3,4-dihydro-1,2-diazete 1,2-dioxides and their vasodilatory properties in vitro and in vivo. *Bioch.Biophys.Res.Comm.* **214**: 1023-1032.

Weiner, L.M., Hu, H., and Swartz, H.M. 1991, EPR method for the measurement of cellular sulfhydryl groups. *FEBS Letters* **290**: 243-246.

Woldman, Ya.Yu., Khramtsov,V.V., Grigor'ev, I.A., Kiriljuk, I.A., and Utepbergenov, D.I., 1994, Spin trapping of nitric oxide by nitronylnitroxides: measurement of activity of NO synthase from rat cerebellum. *Bioch.Bioph.Res.Comm.* **202**: 195-203.

Yelinova, V.I., Glazachev, Y.I., Khramtsov, V.V., Kudryashova, L.N., Rykova, V., and Salganik, R.I., 1996, Studies of human and rat blood under oxidative stress: changes in plasma thiol level, antioxidant enzyme activity, protein carbonyl content and fluidity of erythrocyte membrane. *Biochem. Biophys. Res. Comm.* **221**:300-303.

Yelinova, V.I., Khramtsov, V.V., and Markel, A.L., 1999, Manifestation of Oxidative Stress in the Pathogenesis of Arterial Hypertension in ISIAH Rats. *Biochem. Biophys. Res. Commun.* **263**: 450-453.

Yelinova, V.I., Mazhukin, D.G., and Khramtsov, V.V., 2000, Synthesis of novel nitronylnitroxide and its application for nitric oxide trapping. In *6th International Symposium on Spin Trapping.* Marselle (France), pp. 113.

Yoshida, M., Akaike, T., Wada, Y., Sato, K., Ikeda, K., Ueda, S., and Maeda, H., 1994, Therapeutic effect of imidazolineoxyl N-oxide agains endotoxin shock through its direct nitric oxide-scavenging activity. *Biochem.Biophys.Res. Commun.* **202**: 923-930.

Zamaraev, K.I., Romannikov, V.N., Salganik, R.I., Wlassoff, W.A., and Khramtsov, V.V., 1997, Modelling of Prebiotic Synthesis of Oligopeptides: Silicate Catalysts help to overcome the critical stage. *Origins Life Evol.Biosphere* **27**: 325-337.

New Approaches in Spin Labeling and Spin Trapping. Part Two: NMR Detects Free Radicals

VALERY V. KHRAMTSOV[1,3], LAWRENCE J. BERLINER[2] and THOMAS L. CLANTON[3]
[1]*Institute of Chemical Kinetics and Combustion, Novosibirsk, Russia,* [2]*Department of Chemistry,* [3]*The Ohio State University Heart Lung Institute, Division of Pulmonary and Critical Care Medicine, The Ohio State University, Ohio, USA*

1. INTRODUCTION

The significance of free radical chemistry in biochemical processes is increasingly appreciated. The technology available for detecting free radical reactions improved over the past 30 years and now includes a large variety of methods encompassing detection of chemical reaction products and fluorescent probes. However, none are as specific as electron spin resonance (ESR), particularly when coupled with spin trapping technology, which remains the gold standard for verification of the presence of free radical species. Furthermore, ESR still has the greatest potential for measurement *in vivo* in real time. Nevertheless, despite the specificity of ESR, it has significant limitations for *in vivo* applications particularly due to rapid reduction of the spin adducts in living cells and in the extracellular fluid to ESR silent products. The other obstacle for *in vivo* ESR particularly in any human application is RF penetration. Therefore, except for a few experiments (Liu *et al.*, 1999; Timmins *et al.*, 1999) *in vivo* measurements of ESR-detectable spin adducts from living systems are usually reported in buffer-perfused organs or cells. In whole animals, sensitivity is generally near the current limit of detection (10-100 μM). Those ESR applications which have proven the most useful have often relied on *ex vivo* samples, using X band.

Recently, we have demonstrated a new approach, termed "NMR spin trapping" (ST-NMR), by following the ^{31}P-NMR spectra of a phosphorus containing nitrone spin trap, DEPMPO, after reaction with a variety of free radical species (Khramtsov *et al.*, 1999). Why was NMR spectroscopy proposed for free radical detection? Normally, the NMR spectrum of a radical is paramagnetically broadened preventing its detection. However, many of the degradation/decomposition reactions of spin traps lead to diamagnetic products and can be studied by NMR. Proton NMR is simply too rich in resonance lines to easily resolve anything useful; however, two alternate stable isotopes with reasonable sensitivity are ^{31}P and ^{19}F. Using ^{31}P- or ^{19}F-NMR in combination with the spin traps containing these nuclei should eliminate the multitude of overlapping NMR signals that would occur with the more common ^{1}H or ^{13}C nuclei and, therefore, potentially provide some valuable information about spin adducts and their degradation reactions. ST-NMR retains the most important feature of ESR spin trapping, namely sensitivity to the structure of the trapped radical. The main limitation of the method, namely the reduced sensitivity of NMR versus ESR, is overcomed by accumulating the stable diamagnetic products formed during the spin adducts degradation. This stability of the diamagnetic products in the reducing environment gives ST-NMR its most important advantage over ESR in biological systems.

In this paper we review recent data on the development and application of ST-NMR technology, its advantages and limitations compared with traditional ESR spin trapping.

2. ST-NMR USING THE PHOSPHORUS-CONTAINING NITRONE SPIN TRAP, DEPMPO

The products from co-incubation of 5-diethoxyphosphoryl-5-methyl-1-pyrroline-N-oxide, DEPMPO, with free radical generators result in unique ^{19}F-NMR spectra that reflect specific free radicals and are particularly valuable for evaluating secondary degradation reactions of the radical adducts. A summary of the stable NMR peaks detected in response to incubation of DEPMPO with representative *in vitro* oxygen and carbon centered radicals are listed in Table 1. Importantly, as we have shown earlier (Khramtsov *et al.*, 1999) these ^{31}P-resonances represent only diamagnetic products of the DEPMPO/adduct degradation. The chemical shifts of the products of the decay for DEPMPO adducts with $^{\bullet}OH$, $^{\bullet}OOH$, $^{\bullet}CH_2OH$, and $^{\bullet}CH_3$ radicals are in the range from 27 to 33 ppm which is far from the range of main phosphorus-containing compounds in biological samples (from -3 to

21 ppm for ATP, ADP, AMP, NAD, inorganic phosphate, creatine phosphate, etc.).

Table 1. Primary NMR peaks[a] for DEPMPO (^{31}P) and FDMPO (^{19}F) reaction products

Reactions	Chemical shifts (± 0.02 ppm)				
DEPMPO only	23.7				
DEPMPO + $^{\bullet}OH$	23.7	27.05			
DEPMPO + $O_2^{\bullet-}$	23.7	27.05			
DEPMPO + $^{\bullet}CH_3$	23.7		24.54	30.83	32.31
DEPMPO + $^{\bullet}CH_2OH$	23.7		24.05	30.80	32.52
FDMPO only	-66.0				
FDMPO + $^{\bullet}CH_3$	-66.0	-78.6[b]	-80[b]		
FDMPO + $^{\bullet}CH_2OH$	-66.0	–74.6[b]	–76.75[b]		

[a]relative to external 85% orthophosphoric acid and to external 1% $CFCl_3$ in d_6-acetone for ^{31}P- and ^{19}F-NMR peaks, respectively.
[b]the peaks appeared only after addition of 10 mM ascorbate.

There are a number of advantages to monitoring only diamagnetic reduction products and not the radical adducts, especially since these diamagnetic products are stable over long time periods. First, the detection of comparatively stable diamagnetic products of degradation of paramagnetic spin adduct allows accumulation of product into the NMR sensitivity range at a point in time that the ESR detectable signal has undergone decay. Figure 1 demonstrates that at high radical concentrations the NMR-detected diamagnetic product is larger than the ESR detected DEPMPO/radical adduct by approximately two orders of magnitude.

The ability to accumulate diamagnetic products of spin adduct degradation significantly decreases the gap in sensitivity between NMR and ESR. Moreover, this may be particularly important for *in vivo* experiments, where the paramagnetic adducts are even more unstable, since cell systems have a high capacity for bioreduction (Swartz *et al.*, 1995a, 1995b). Figure 2a demonstrates this advantage of ST-NMR in highly reducing medium (in the presence of ascorbate) under conditions “simulating” *in vivo* permanent radical product generation via a "recycling" of Fenton chemistry and steady accumulation of product over time. NMR data support the formation of DEPMPO adducts with C-centered radicals during the cyclical Fenton-type oxidation driven by reducing agent (ascorbate) in the presence of 10% (v/v) DMSO (Fig.2a) or MeOH (see Khramtsov *et al.*, 1999). Note that ESR spin trapping can not be applied at these conditions due to fast reduction of nitroxide moiety (less then 1 sec).

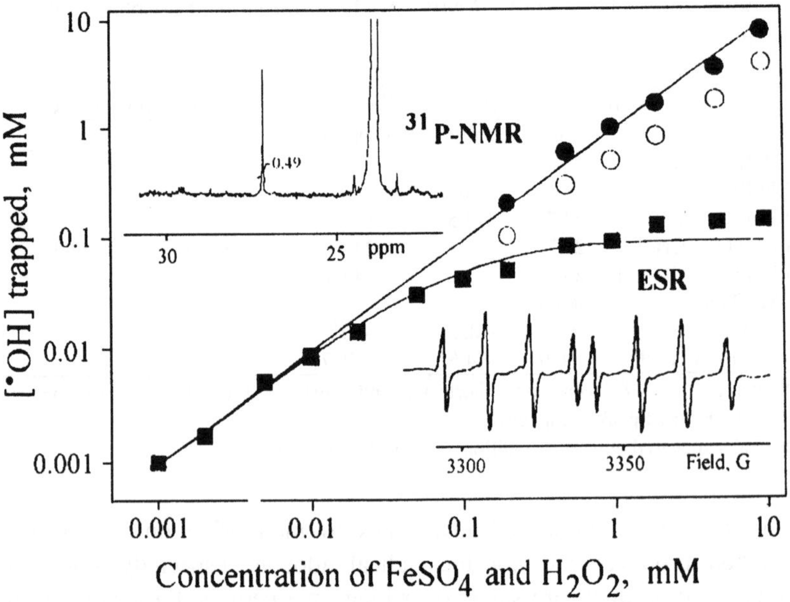

Figure 1. The concentration dependence of DEPMPO/•OH reaction products from the double integrated ESR spectrum (filled squares) and from the integral intensity of the ^{31}P-NMR peak at 27.05 ppm (open circles). The spectra of DEPMPO reaction products were measured 10 min (ESR) and 2h (^{31}P-NMR) after initiation of the Fenton reaction and are shown for 1 mM concentrations of the Fenton reactants. The bent curve was calculated proposing bimolecular decay of DEPMPO/•OH adduct while the straight line is calculated from a 1:1 stoichiometry between the concentration of Fenton reagents and adducts measured. Filled circles represent the same integral intensities of the NMR peak at 27.05 ppm multiplied by factor 2.0. This reflects the transformation of 50 % of initial DEPMPO/•OH adduct first in the corresponding hydroxylamine upon its bimolecular degradation, and then back into the parent nitrone, DEPMPO (see Scheme 2).

3. ST-NMR USING THE FLUORINE-CONTAINING NITRONE SPIN TRAP, FDMPO

The main obstacle for NMR spin trapping applications in biological systems currently seems to be comparatively low sensitivity of NMR spectroscopy even taking into account the possibility of adduct accumulation. Fluorine-containing spin traps would theoretically show at least an order of magnitude higher sensitivities compared with ^{31}P-containing traps. Selinsky *et al.* (1989) proposed a series of fluorinated PBN (N-tert-butyl-phenylnitrone) analogs as NMR active spin traps and demon-

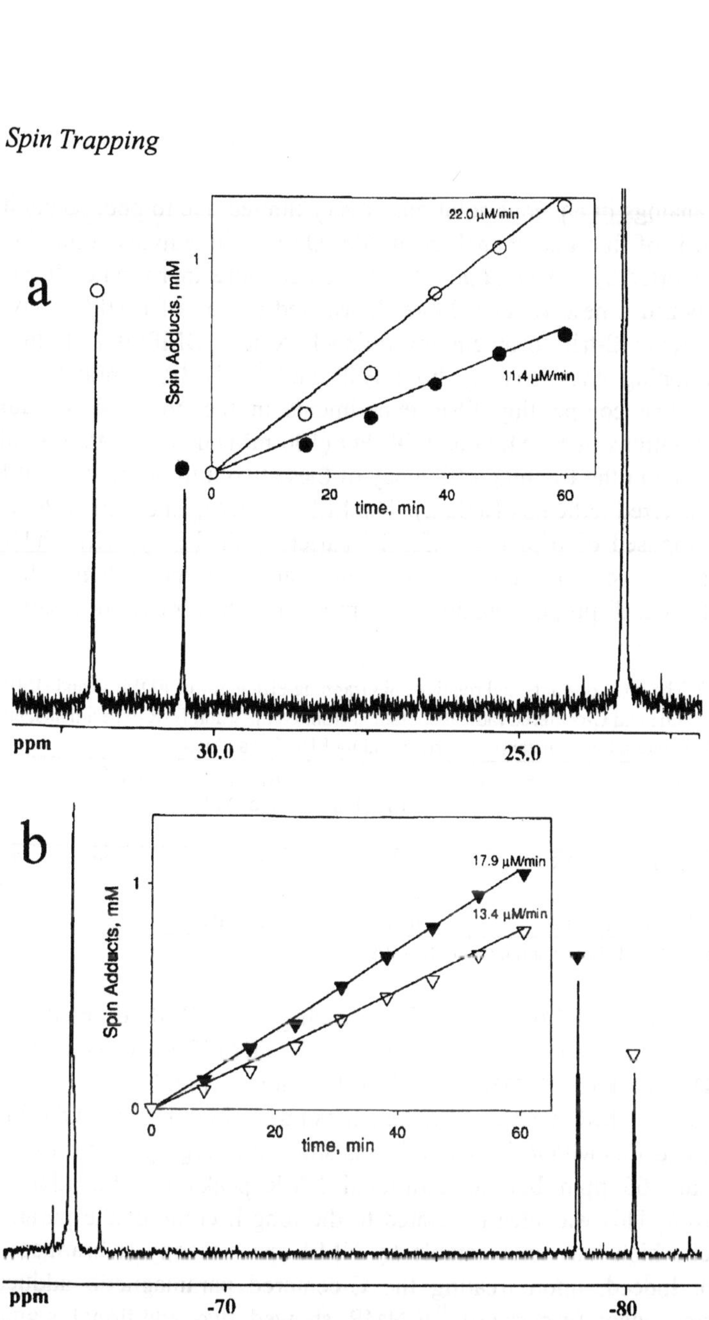

Figure 2. ^{31}P-NMR spectrum of reduced $^{\bullet}CH_3$ adduct of DEPMPO (a) and ^{19}F-NMR spectrum of FDMPO (b) in a Fenton "iron recycling" system in 0.1 M K-phosphate buffer, pH 7.0, 2 mM DTPA, 20 mM H_2O_2, 50 mM spin trap, 10% v/v DMSO, 20 mM ascorbate, 30 min after addition of 0.5 mM $FeSO_4$ solution; number of scans 208; inset: kinetics of accumulation of the products calculated from the peaks that are denoted by corresponding symbols.

strated their trapping efficiency in the thermal decomposition of phenylazotriphenylmethane using ^{19}F-NMR. Since the use of fluorinated

PBN analogs in aqueous solutions is very limited due to poor solubility, the synthesis of fluorinated analogs of DMPO (5,5'-dimethylpyrroline-1-oxide), first reported by Janzen *et al.* (1995), seemed more appropriate. Recently we synthesized a new water soluble fluorinated analog of DMPO, 4-hydroxy-5,5-dimethyl-2-trifluoromethylpyrroline-1-oxide (FDMPO) and studied its spin-trapping chemistry by both ESR and ^{19}F-NMR (Khramtsov *et al.*, 2000). The competition ESR experiments in the presence of equimolar concentrations of FDMPO and DMPO (100 mM) demonstrate insignificant difference in the trapping efficiency of these two nitrone traps towards OH- or C-centered radicals (Table 2). The ESR spectra of these FDMPO adducts are composed of triplet of 1:3:3:1 quartets with the hyperfine interaction constants given in Table 2, and were stable for more than 8 hrs since bimolecular disproportionation is suppressed in the absence of a α-proton.

Table 2. The rate constants and relative spin trapping efficiency of FDMPO and DMPO towards $^{\bullet}OH$ and C-centered radicals obtained from competition ESR experiments, and hyperfine interaction constants of corresponding FDMPO adducts.

Spin adduct	k_{FDMPO}/k_{DMPO}	k_{DMPO} x10^8 $M^{-1}s^{-1}$	$K_{FDMPO,}$ x10^8 $M^{-1}s^{-1}$	a_N, G	a_F, G
FDMPO/$^{\bullet}$OH	0.8	28^a	22.4±5	13.90	2.75
FDMPO/$^{\bullet}CH_3$	0.65	0.14^b	0.1±0.03	14.90	2.05
FDMPO/$^{\bullet}CH_2OH$	0.9	0.22^b	0.2±0.04	14.75	2.4

[a] Madden et al, 1996; [b]Taniguchi et al., 1999

The parent FDMPO trap exhibits a single ^{19}F-NMR resonance at –66.0 ppm. The relative signal to noise ratio of the FDMPO signal compared to the ^{31}P-NMR peak of DEPMPO was 10.4 fold higher (Fig. 3).

The ^{19}F-NMR spectra of the reactions producing FDMPO/$^{\bullet}$OH, FDMPO/$^{\bullet}CH_3$, and FDPMPO/$^{\bullet}CH_2OH$ adducts showed a large parent nitrone single peak at –66 ppm but no additional NMR peaks for the adducts were observed. This was clearly related to the long lifetime of the paramagnetic adduct which was undetectable by NMR spectroscopy (Khramtsov *et al.*, 1999). Indeed, upon treating the C-centered paramagnetic adducts with reducing agent (ascorbate) ^{19}F-NMR showed two additional signals (see Table 1). Addition of ascorbate to the FDMPO/$^{\bullet}$OH adduct did not result in the appearance of additional NMR peaks. This might be related to the chemical instability of hydroxylamine from the FDMPO/$^{\bullet}$OH adduct, which degrades into the initial parent FDMPO nitrone (see section 4).

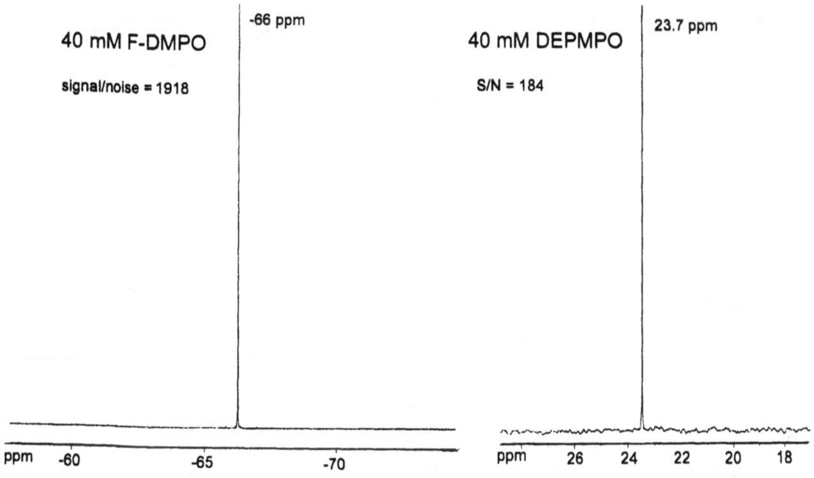

Figure 3. Comparison of NMR sensitivity of FDMO and DEPMPO spin traps. Spectra taken on Bruker DPX-250 spectrometer at 235 MHz (^{19}F) or 101 MHz (^{31}P) in 5 mm tubes in water solution with 10 percent v/v D_2O; number of scans, 16; acquisition time, 1 sec.

We were also able to follow the kinetics of C-centered radical production in the cyclical Fenton-type reaction, which is driven by the reducing agent, ascorbate, in the presence of 10% MeOH or 10% DMSO (Fig.2b). Here ESR cannot be applied, due to the fast reduction of the nitroxide moiety. The results depicted for FDMPO/$^{\bullet}CH_3$ adduct (Fig.2b) show two resonances at -78.6 ppm and -80 ppm increasing with time in parallel. The same two peaks at -78.6 ppm and -80 ppm were observed for the synthesized hydroxylamine (HA; see Scheme 1 for the structure, where X=OH, Y=R=CH_3, Z=CF_3). Moreover, the ^{19}F spectrum of synthesized stable radical, FDMPO/$^{\bullet}CH_3$, does not show any NMR signals but only reveals the two peaks, again at -78.6 ppm and -80 ppm, after ascorbate addition (data not shown).

4. SPIN ADDUCT PRODUCTS MEASURED BY ST-NMR

In principle, the ability to distinguish and quantify multiple degradation products by NMR, also provides insight into the specific degradation pathways which can be extremely useful and may be unique to particular *in vivo* environments. In the case of nitrone spin traps, radical adducts R1 and R2 are formed according to the Scheme 1 (X=Z=H, Y=OP(OEt)$_2$ for DEPMPO and X=OH, Y=Me, Z=CF_3 for FDMPO):

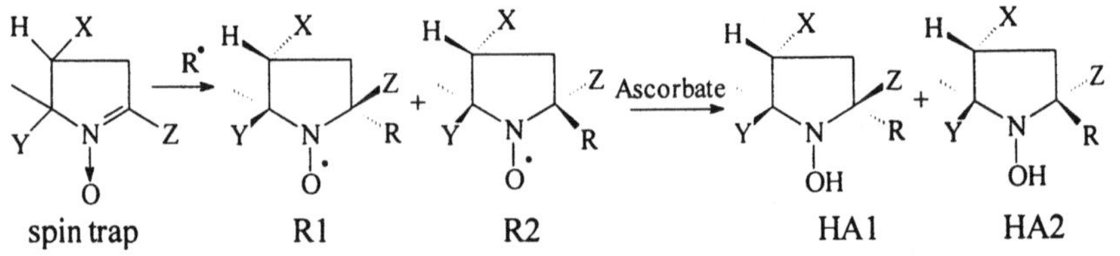

Scheme 1.

Paramagnetic adducts R1 and R2 are two stereoisomers that in some cases can be distinguished by ESR, e.g. for the DEPMPO/$O_2^{\bullet -}$ or DEPMPO/$^\bullet CH_2OH$ adducts (Frejaville *et al.*, 1995). NMR does not detect radical adducts due to paramagnetically broadened spectra but diamagnetic products of the radical adducts degradation. Indeed, in the presence of reducing agent (ascorbate) two additional NMR peaks were observed using both DEPMPO and FDMPO for the C-centered radical generating system (Fig.2). Moreover, the total rates of adduct accumulation were almost identical (31.6 and 31.3 μM/min, respectively) for the DEPMPO and FDMPO radical adducts in the "iron-recycling" system measured by ^{31}P- and ^{19}F -NMR, respectively. These two peaks were assigned to the two stereoisomers of the corresponding diamagnetic hydroxylamines, HA1 and HA2, formed upon the reduction of the paramagnetic adducts (see Scheme 1). This was proved for FDMPO/$^\bullet CH_3$ adducts by synthesis of the hydroxylamines, HA1 and HA2 (X=OH, Y=R=Me, Z=CF_3), as an analytically pure sample (Khramtsov *et al.*, 2000). The ^{19}F-NMR spectrum of these compounds demonstrates the same two identical resonances as shown in Fig. 2b.

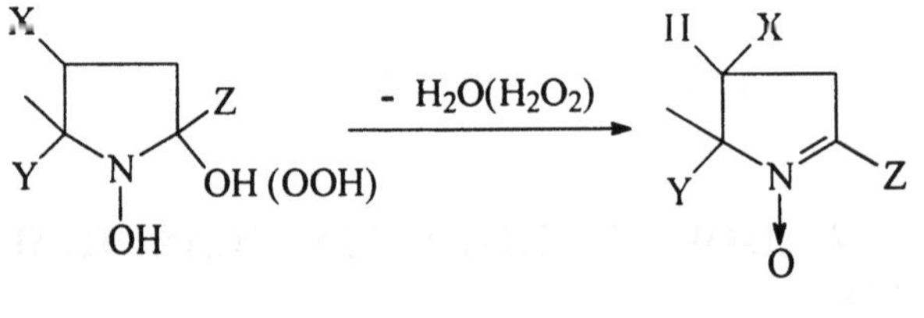

Scheme 2.

We have encountered some practical problems to date with ST-NMR detection of oxygen-centered radicals. For example, we did not observe any additional signal resulting from reaction of DEPMPO or FDMPO with either hydroxyl or superoxide radicals in the presence of ascorbate, which suggests some additional conversion of the hydroxylamines, HA1 and HA2 (Scheme 1). This conversion results in regeneration of the parent nitrones according to the Scheme 2 (Khramtsov *et al.*, 1999):

Interestingly, the proposed mechanism can provide some insight into the antioxidant properties of these spin probes, which may have medical significance (Pietri *et al.*, 1998).

In the absence of strongly reducing agents we have observed the same ^{31}P-line (at 27.05 ppm, see Table 1) resulting from reaction of DEPMPO with either hydroxyl or superoxide radicals which was assigned to the product of oxidation of the radical adducts (Khramtsov *et al.*, 1999). The observation by NMR of only 50% of the diamagnetic product formed from DEPMPO/$^{\bullet}$OH in Fenton system during adduct bimolecular decay supports backward transformation of the reduced product, hydroxylamine, into the parent nitrone (Fig. 1).

5. CONCLUSION

This report reviews a new method of detecting free radicals using ^{31}P- or ^{19}F-NMR. We envision this approach leading to development of new spin trapping agents, specifically designed for NMR, and their possible utilization as a diagnostic tool for imaging of areas of inflammation and free radical production *in vivo* in man, where the penetration and depth limitations of ESR become problematic. Although ST-NMR approach is relatively in its infancy, it shows promise of improving our ability to someday detect free radicals production in real time and possibly "imaging" oxidative stress in living organisms in a quantitative way.

The main potential obstacle with the ST-NMR approach is still the comparatively low sensitivity of NMR spectroscopy, even taking into account product accumulation. Nevertheless Delmas-Beauvieux (1997) recently reported a characteristic NMR peak, shifted 3.3 ppm down from the parent DEPMPO, taken 1-hour after warm ischemia and reperfusion of isolated rat liver. The identity of this peak with the diamagnetic products of DEPMPO-$^{\bullet}$OH degradation (Table 1), supports the observation of free radical generation in this ischemic injury model and, therefore, the potential use of NMR spin trapping in biomedicine.

As to future signs of NMR sensitive compounds for free radical detection, note that these do not actually have to be spin traps at all, but rather compounds that react with high fidelity and redox specificity in a biological environment. These might be fluorinated analogs of antioxidants such as the ones used for investigation of the antioxidant capacity of biomolecules and biofluids (Aime et al., 1999), SOD-mimetics influencing the relaxation of F-containing compounds (Rigo *et al.*, 1981, 1985) or other redox sensitive compounds containing ^{19}F or ^{31}P nuclei.

ACKNOWLEDGMENTS

The authors wish to acknowledge funding for this research in part from the grants of from USPHS (GM58772) and Russian Found.Basic Res. (99-0449921).

REFERENCES

Aime, S., Calzoni, S., Digilio, G., Giraudo, S., Fasano, M., and Maffeo, D., 1999, A Novel ^{19}F-NMR Method for the investigation of the antioxidant capacity of biomolecules and biofluids. *Free Rad .Biol. Med.* **27**: 356-363.

Delmas-Beauvieux, M. C., Pietri, S., Culcasi, M., Leducq, N., Valeins, H., Liebgott, T., Diolez, P., Canioni, P., and Gallis, J. L., 1997, Use of spin-traps during warm ischemia-reperfusion in rat liver: comparative effect on energetic metabolism studied using ^{31}P nuclear magnetic resonance. *Magma* **5**: 45-52.

Frejaville, C., Hakim, K., Tuccio, B., Le Moigne, F., Culcasi, M., Pietri, S., Lauricella, R., and Tordo, P., 1995, 5-(Diethoxyphosphoryl)-5-methyl-1-pyrroline -oxide: a new efficient phosphorylated nitrone for the *in vitro* and *in vivo* spin trapping of oxygen-centered radicals. *J. Med. Chem.* **38**: 258-265.

Janzen, E.G., Zhang, Y-K., and Arimura, M., 1995, Synthesis and spin-trapping chemistry of 5,5-dimethyl-2-(trifluoromethyl)-1-pyrroline N-oxide. *J .Org. Chem.* **60**: 5434-5440.

Khramtsov, V.V., Berliner, L.J., Clanton. T.L., 1999, NMR spin trapping of free radical reactions using a phosphorus-containing nitrone spin trap. *Magn. Reson Med.* **42**: 228-234.

Khramtsov, V.V., Litkin, A.K., Reznikov, V.A., Berliner, L.J., and Clanton, T.L., 2000, Comparative NMR spin trapping studies of phosphorus and fluorine containing analogs of DMPO. In *6th International Symposium on Spin Trapping.* Marselle (France), pp.68.

Liu, K.J., Miyake, M., Panz, T., and Swartz, H.M., 1999, Evaluation of DEPMPO as a spin trapping agent in biological systems. *Free Rad. Biol. Med.* **26**: 714-721.

Madden, K.P., Taniguchi, H.J., 1996, In situ radiolysis time-resolved ESR studies of spin trapping by DMPO: reevaluation of hydroxyl radical and hydrated electron trapping rates and spin adduct yields. *J .Phys. Chem.* **100**: 7511-7516.

Pietri, S., Liebgott, T., Frejaville, C., Tordo, P., and Culcasi, M., 1998, Nitrone spin traps and their pyrrolidine analogs in myocardial reperfusion injury: hemodynamic and ESR implications. *Eur. J. Biochem.* **254**: 256-265.

Rigo, A., Viglino, P., and Rotilio, G., 1985, NMR assay based on ^{19}F. In *Handbook of Methods for Oxygen Radical Research* (R.A. Greenwald, Ed.) CRC Press, Boca Raton, FL, pp.227-232.

Rigo, A., Ugo, P., Viglino, P., and Rotilio, G., 1981, ^{19}F nuclear magnetic relaxation by superoxide dismutase as an enzymatic method for detection of superoxide ion. *FEBS Lett.* **132**: 78-80.

Selinsky, B.S., Levy, L.A., Motten, A.G., and London, R.E., 1989, Development of fluorinated, NMR-active spin traps for studies of free radical chemistry. *J Magn. Reson.* **81**: 57-67.

Swartz, H.M., Sentjurc, M., and Kocherginsky, N., 1995a, Metabolism of nitroxides and their products in cells. In *Nitroxide Spin Labels: Reactions in Biology and Chemistry* (N.Kocherginsky, and H.M Swartz, eds.) CRC Press, Boca Raton, pp.113-147.

Swartz, H.M., Sentjurc, M., and Kocherginsky, N., 1995b, Metabolism and distribution of nitroxides in vivo. In *Nitroxide Spin Labels: Reactions in Biology and Chemistry* (N.Kocherginsky, and H.M Swartz, eds.) CRC Press, Boca Raton, pp.153-173.

Taniguchi, H.J., and Madden, K.P., 1999, An *in situ* radiolysis time-resolved ESR study of the kinetics of spin trapping by 5,5-dimethyl-1-pyrroline-N-oxide. *J. Am .Chem. Soc.* **121**: 11875-11879.

Timmins, G.S., Liu, K.J., Bechara, E.J.G., Kotake, Y., and Swartz, H.M., 1999, Trapping of free radicals with direct *in vivo* EPR detection: a comparison of 5,5-dimethyl-1-pyrroline-N-oxide and 5-diethoxyphosphoryl-5-methyl-1-pyrroline-N-oxide as spin traps for HO and $SO_4^{\cdot-}$. *Free Rad. Biol. Med.* **27**: 329-333.

Free Radical Intermediates of Drugs and Xenobiotics

LAWRENCE J. BERLINER AND HIROTADA FUJII
Department of Chemistry, The Ohio State University, Columbus, Ohio 43210, USA and Sapporo Medical University, Sapporo, Hokkaido, Japan

1. INTRODUCTION

Although the levels of "naturally occurring" free radicals, such as NO and flavins, semiquinones and the like are present at quite low levels they are, in fact, essential for life. We are generally unaware of other free radical metabolism that occurs in biology with the exception of reactive oxygen species (ROS) such as superoxide, hydroxyl radical and a variety of lipoxy radicals, which are quite short-lived and result from severe circumstances such as ischemia-reperfusion or hyperoxia. While the consequences of a burst of ROS are well appreciated and are the focus of an intense effort in antioxidant research, there may be several classes of more stable free radicals lurking in tissues and cells.

In particular, in occupations where high levels of organic or petrochemicals are ingested, the radical chemistry has been mostly ignored in pharmacology-toxicology studies. In addition, we voluntarily ingest large quantities of organic pharmaceuticals that are intended to alleviate a variety of disease symptoms. This chapter presents two examples of xenobiotics that result in persistent radical intermediates in vivo.

2. NITROSOBENZENE

This compound is toxic and is labeled as a potent mutagen. Hence, why would one want to study its biological fate? On the one hand, we may not be able to avoid exposure to small trace levels. On the other hand, Figure 1 depicts the various oxidation states of nitrogen substituted aromatics going

Supramolecular Structure and Function 7, edited by Pifat-Mrzljak.
Kluwer Academic/Plenum Publishers, New York, 2001.

from aniline through to nitrobenzene. It is important to point out that there are hepatic and intestinal flora enzymes capable of amine oxidase and nitroreductase activity which can, in principle, convert anilines and nitrobenzenes to the more potent nitroso form. Our interests focus on two aspects: radical forms of nitrosobenzene and the in vivo chemistry of this compound.

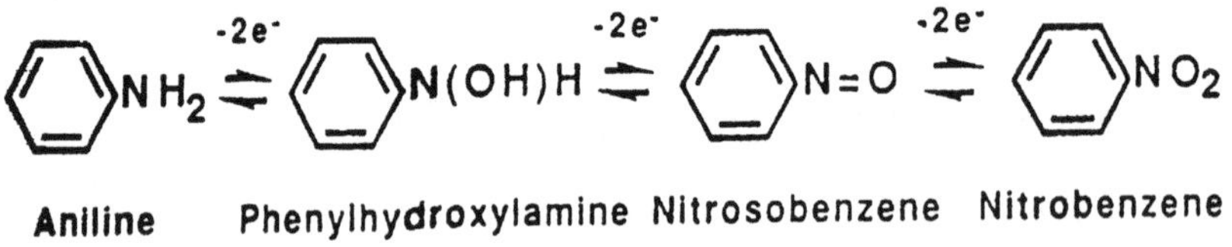

Figure 1. Oxidation states of nitrogen substituted aromatics

2.1 Phenylhydronitroxide radical

Figure 2 depicts typical X-band spectra of nitrosobenzene in hemolyzed blood, or in pure mixtures of nitrosobenzene and NADH or vitamin C (Fujii *et al.*, 1994a). However, this spectrum rapidly decays unless a continuous, stoichiometric supply of oxygen is provided, since addition of glucose oxidase scavenged all of the O_2, resulting in the disappearance of the EPR spectrum. This complex, multiline spectrum can be simulated as the phenylhydronitroxide radical, which results from a one-electron reduction of nitrosobenzene. The detailed hyperfine structure results from a distribution of spin density at the various non-equivalent protons on the ring. Since most redox chemistry is two electron, this radical results from either the one electron "back oxidation" of the hydroxylamine by oxygen (yielding a molecule of superoxide) or a disproportionation reaction between a nitrosobenzene and nitrosobenzene-phenylhydroxylamine molecule, yielding two phenylhydronitroxides. The overall scheme is shown in Figure 3. Of course, the pressing question is: "what happens in-vivo?"

2.2 In vivo EPR of nitrosobenzene adducts

Figure 4a depicts an L-band in vivo EPR spectrum of a 15-20 g mouse after injection with an ethanol or liposome suspension of nitrosobenzene. Note the broad three line spectral component, which reflects an immobilized

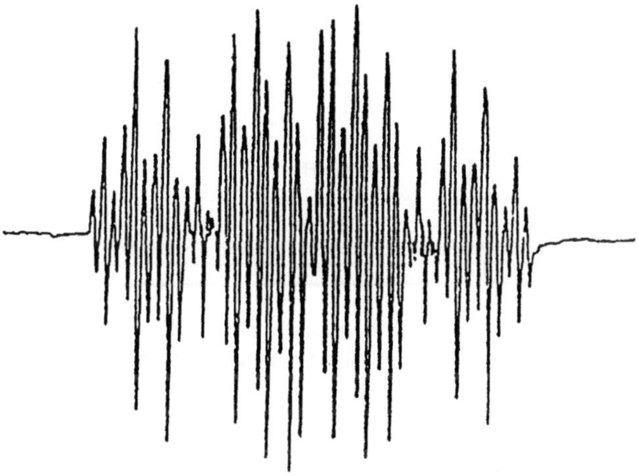

Figure 2. Typical X-band EPR spectra of the reaction of nitrosobenzene and NADH. Effects of glucose oxidase on the nitrosobenzene/NADH model system. The mixture contains 2mM nitrosobenzene, 1mM NADH, 20 mM glucose (50 mM phosphate buffer, pH 7.4, 25C). The same spectrum was observed for nitrosobenzene in hemolyzed blood. However, the continuous presence of oxygen was required, otherwise the spectrum slowly decreased. This was confirmed by an experiment in the presence of 2mM nitrosobenzene, 1mM NADH, 20 mM glucose plus glucose oxidase (40 units/ml) where the spectrum was nonexistent since glucose oxidase totally consumes the oxygen in the sample. Instrumental conditions were: microwave power, 5 mW; modulation amplitude, 0.1 gauss; time constant, 0.1 sec; scan rate, 50; gauss/min.

nitroxyl radical, which is *not* a slowly tumbling phenylhydronitroxyl radical as seen in Figure 2 (Fujii *et al.*, 1994b). As a control, an animal was injected with the vehicle alone (liposome suspension or ethanol) to show that the observed spectrum did not arise from artifacts of the injection protocol (Figure 4b).

In an alternate approach, we excised buttocks muscle tissue from an anesthetized mouse within 5 minutes of injection and observed in an X-band EPR tissue cell (Figure 5). Note how the multiline phenylhydronitroxide radical spectral component disappears while the broad three-line nitroxyl component grows and remains constant for several hours (in fact our in vivo studies showed that this broad-line EPR spectrum remained even days after death!). We can rationalize the initial multiline spectral component since hemolysis occurs during tissue excision; however, the persistent radical spectrum was a mystery!

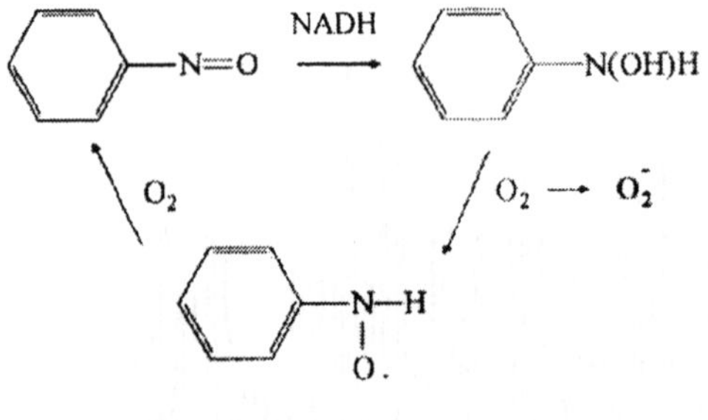

Figure 3. Reduction pathway of nitrosobenzene

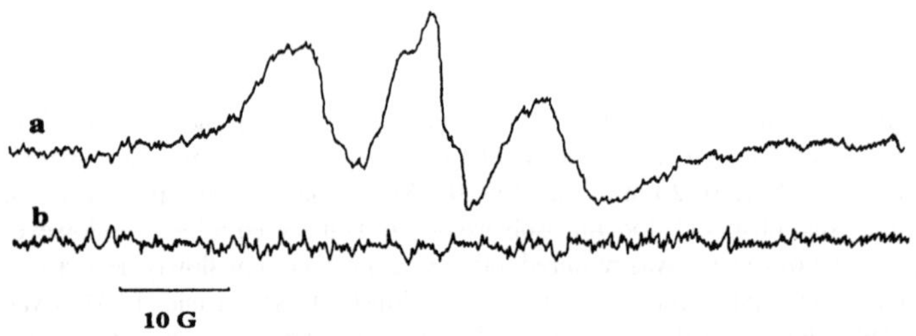

Figure 4. L-Band EPR spectra of an anesthetized (25g) mouse at room temperature. (a) 20 min. after an ip injection of 0.1ml of 50mM nitrosobenzene (in a liposome suspension); (b) control injection without nitrosobenzene. Spectrometer conditions: frequency, 1.128GHz; microwave power, 10mW; modulation, 1.9G; applied magnetic field, 390G; sweep width, 100G; sweep rate, 25G/min.; time constant, 1.0. From Fujii et al. (1994b) with permission.

After observing this excised muscle tissue at higher microwave power and comparing it with the reaction product from mixing nitrosobenzene with oleic acid or other unsaturated fatty acid (Figure 6) it was clear that the chemistry that occurred in muscle tissue was spectrally identical with that from a pure oleic acid sample. (Note that a saturated fatty acid, such as stearic acid, does not react). Figure 7 depicts the mechanism of "addition" of nitrosobenzene across a double bond, termed either as an "ene-addition" or a pseudo Diels-Alder reaction (Sullivan. 1966), resulting in a migration of the double bond, and a hydroxylamine intermediate which is easily oxidized to the nitroxyl adduct (nitroxide I) by mechanisms analogous to those shown earlier in Figure 3. In order to confirm that this reaction mechanism made

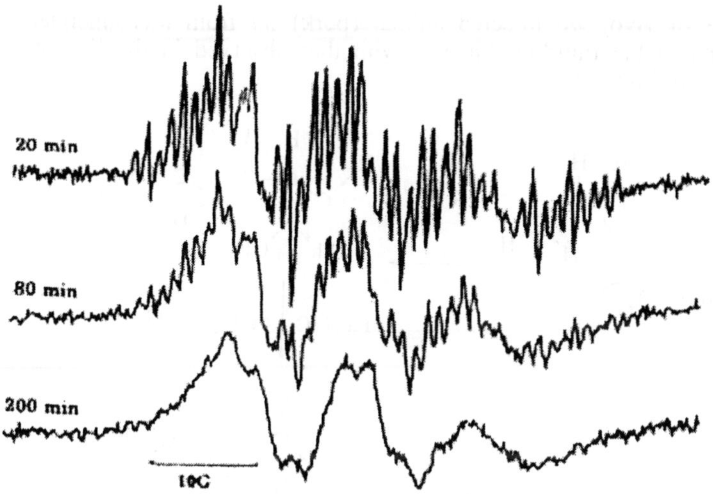

Figure 5. X-Band EPR spectra of excised muscle tissue after in-vivo i.m. administration of 0.1ml 50mM nitrosobenzene (in ethanol) in the mouse buttocks: (a) 20 min. or (b) 80 min. after injection, and (c) 200 min. after injection. The muscle tissue was removed within 5 minutes after injection. Spectrometer conditions were: microwave power, 10mW; frequency, 9.25GHz; modulation, 0.5G; applied magnetic field, 3400G; sweep width, 100G; sweep rate, 25G/min.; time constant, 0.128s. From Fujii *et al.*, (1994b) with permission.

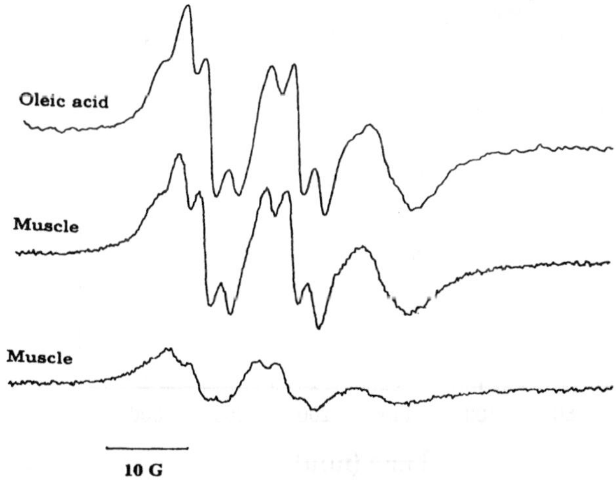

Figure 6. Comparison of the reaction product of nitrosobenzene and oleic acid with excised muscle tissue at high gain.

sense in vivo, we injected animal (pork) fat from the supermarket and compared the build-up kinetics with that observed in the live mouse, as shown in Figure 8.

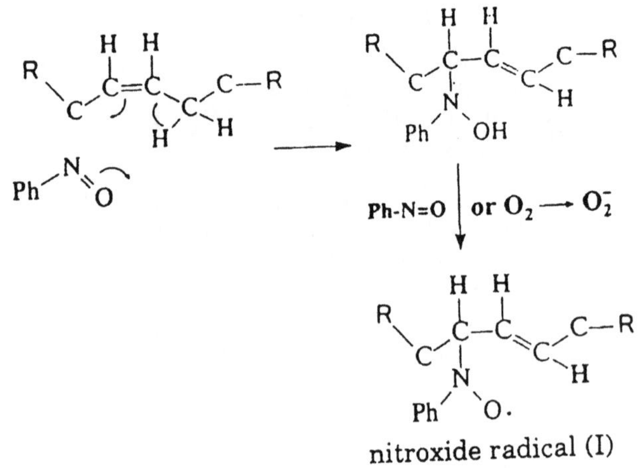

Figure 7. Pseudo Diels-Alder "ene-addition" reaction scheme.

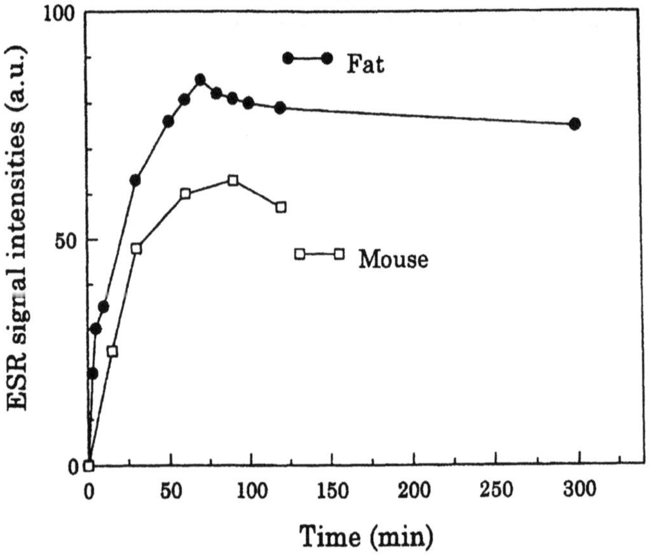

Figure 8. Build-up kinetics of pure fat with that observed in the live mouse. The same bolus of nitrosobenzene used in the experiments depicted in Figures 4 and 5 was injected into a piece of pork fat.

In summary, nitrosobenzene can form persistent radical adducts with unsaturated fatty acids in lipophilic tissue, where it remains inaccessible to most bioreductants, resulting in a strong, stable nitroxyl EPR spectrum. Since there are normal metabolic pathways for converting anilino- and nitroaromatics to the nitroso analogue (see Figure 1), one may speculate that chemical workers continuously exposed to compounds of this type may have stable nitroxyl radical adducts in their fatty tissue. A further concern is that these adducts have been shown to a viable catalysts of lipid peroxidation (Sammartano and Malejka-Giganti, 1991). Both the nitroxyl radical and its hydroxylamine are potentially capable of reducing trace ferric to ferrous ion with possible initiation of lipid peroxidation (Minotti and Aust, 1987).

3. NIFEDIPENE – AN ANTIHYPERTENSIVE PHARMACEUTICAL

Nifedipene (3,5-pyridinedicarboxylic acid-1,4-dihyrdo-2,6-dimethyl-4-(2-nitrophenyl)-dimethylester) is a very popular pharmacological agent known as a slow calcium channel blocker or calcium ion antagonist. It inhibits the transmembrane influx of Ca(II) into cardiac and smooth muscle, inhibiting calcium ion flux across the cell membrane without altering serum calcium concentrations (Lamping and Gross, 1985).

As shown in Figure 9 nifedipene can be photochemically converted to its nitroso-form under high power illumination (Misik *et al.*, 1991; Stasko *et al.*, 1994). There are coincidentally reports that extended sun exposure may

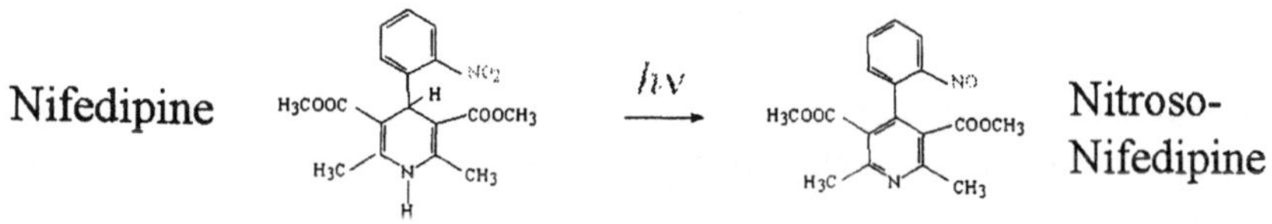

Figure 9. Nifedipene photochemically converted to its nitroso-form.

result in severe skin inflammation and other adverse reactions (Thomas and Wood, 1986). Upon reflecting on the earlier in vivo studies with nitrosobenzene, it seemed prudent to monitor the in vivo EPR of animals exposed to nitrosonifedipine. A second concern was whether low, room light levels of illumination were sufficient to promote the conversion of nifedipine to nitrosonifedipine, especially since Stasko *et al.* (1994) demonstrated nitroxyl adducts with oleic acid similar to that shown earlier in this chapter. If the patient leaves a few nifedipine capsules exposed to room light for a day or two, are there risks?

3.1 Low level light illumination of nifedipine

A 100 mM nifedipine solution (in DMSO) was illuminated under a 15 W household fluorescent lamp for 24 hr. Then a 100 μL bolus was injected i.m. in a 15g anesthetized mouse. The animal was placed in the loop gap resonator localized at the buttocks muscle area 15 minutes later. As shown in Figure 10, a broad three-line spectrum was observed with a 16.8 ± 0.4 gauss nitrogen hyperfine coupling from an immobilized nitroxyl radical adduct, which persisted for more than 1 hr. If tissue was excised 15 min after injection and observed at X-band, an even stronger signal to noise signal was observed for buttocks muscle and especially liver tissue which is also shown in Figure 10.

3.2 Hepatic conversion of nifedipine

Since the liver contains nitroreductases, could the conversion of nifedipine to its nitroso analogue occur without illumination? We soaked freshly excised liver with nifedipine solution that was shielded from direct illumination. Figures 11a and b compare resultant X-band EPR spectra of livers treated with illuminated and non-illuminated nifedipine, respectively. It is clear that the two spectra are superimposable, suggesting that nifedipine was converted by hepatic enzymes to its nitroso analogue. The relative levels of radical adduct was 43 ± 5 and 4.5 ± 0.7 nmol/g of liver in the illuminated and non illuminated samples, respectively. Similar results were observed with nisoldipine, a structural analogue of nifedipine. On the other hand, nifedipine isomer/analogies which are meta-substituted (i.e. nimodipine, nicardipine) do not show radical adducts when illuminated and appear to be, in general, much less susceptible to conversion to the nitrosoform.

4. PHYSIOLOGICAL IMPLICATIONS

There are several subsequent chemical processes associated with nitrosoaromatic chemistry that lend concern from a health standpoint. We showed in previous work that nitrosocompounds can be converted to the hydroxylamine in vivo by various reductants, such as NAD(P)H (Fujii *et al*,

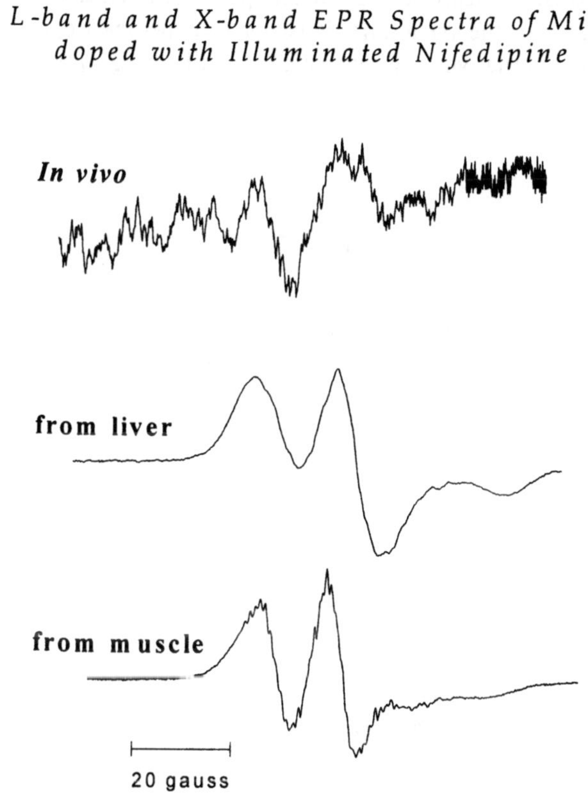

Figure 10. L-band in-vivo EPR spectra of a (15g) mouse at room temperature 15 min after im injection of 100 mM illuminated nifedipine (100 μL in DMSO). Conditions were as in Figure 4. X-band ex-vivo EPR spectra of excised liver tissue and buttocks muscle. Conditions were as in Figure 5. Nifedipine solutions were illuminated by placing solutions in a 5 mL (8 mm i.d.) glass test tube on a 15 W common household fluorescent tube for 24 hrs.

1994a). When reoxidized by oxygen, superoxide is produced stoichiometrically. Furthermore, it has been shown that nitroxyl radical adducts of unsaturated fatty acids can support lipid peroxidation (Sammartano and Malejka-Giganti, 1991). Both the radical and the hydroxyl-amine are potentially capable of reducing trace Fe(III) to Fe(II) with possible initiation of lipid peroxidation (Minotti and Aust, 1987). Lastly, epidemiological studies with patients on nifedipine or nisoldipine concluded that they have 40-60% higher susceptibility to myocardial infarction (Psaty *et al.*, 1995; Psaty *et al.*, 1997). Considering the many hypotheses and controversies over free radicals, ROS, and lipid peroxidation in heart disease, one might take notice of the radical adducts found above with nifedipine. While we can neither infer nor imply that these radical adducts are the cause of heart damage at this point, it is clear that further research is needed. In the meantime, it is our (nonmedical) opinion that patients consider alternative pharmaceuticals until these questions are answered.

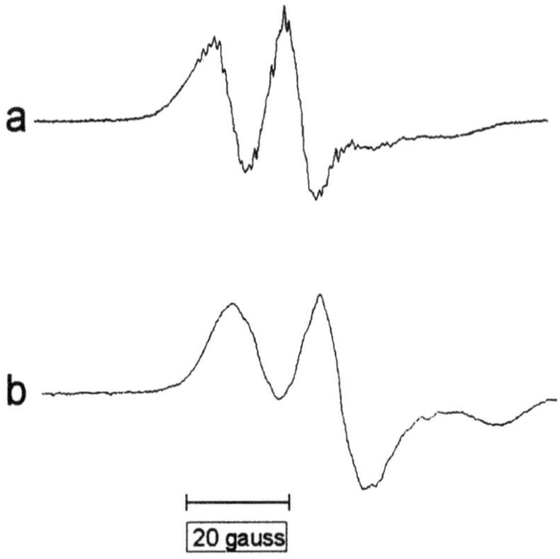

Figure 11. X-band EPR spectra of liver tissue exposed to (a) non-illuminated and (b) illuminated nifedipine. Nifedipine solutions were freshly prepared and soaked with liver tissue for 5 min at room temperature in the dark. Illuminated nifedipine was prepared as described in Figure 10.

REFERENCES

Fujii, H., Koscielniak, J., Kakinuma K., and Berliner, L.J., 1994a, Biological reduction reactions of aromatic nitroso compounds: Evidence for the involvement of superoxide anions. *Free Radical Res. Commun.* **21**: 235-243.

Fujii, H., Zhao, B., Koscielniak, J., and Berliner, L.J., 1994b, First demonstration of in-vivo ESR observation of bioradical formation in a living animal: nitrosobenzene. *Magnetic Reson. Med.* **31**: 77-80.

Lamping, K.A., and Gross, G.J., 1985, Improved recovery of myocardial segment function following a short coronary occlusion in dogs by nicorandil, a potential new antianginal agent, and nifedipine. *J. Cardiovasc. Pharmacol.* **7**: 158-166.

Minotti, G., and Aust, S.D., 1987, The requirement of iron (III) in the initiation of lipid peroxidation by iron II and hydrogen peroxide. *J. Biol. Chem.* **262**: 1098-1104.

Misik, V., Stasko, A., Gergel, D., and Ondrias, K., 1991, Spin-trapping and antioxidant properties of illuminated and nonilluminated nifedipine and nimodipine in heart homogenate and model system. *Molec. Pharmacol.* **40**: 435-439.

Psaty, B.M., Heckbert, S.R., Koepsell, T.D., Siscovick, D.S., Raghunathan, T.E., Weiss, N. S., Rosendaal, F.R., Lemaitre, R.N., Smith, N.L., and Wahl,. P.W., 1995, The risk of myocardial infarction associated with antihypertensive drug therapies. *J. Amer. Med. Assoc.* **274**: 620-625.

Psaty, B.M., Smith, N.L. Siscovick, D.S., Koepsell, T.D., Weiss, N.S., Heckbert, S.R., Lemaitre, R.N., Wagner, E.H., and Furberg, C.D., 1997, Health outcomes associated with antihypertensive therapies used as first-line agents. A systematic review and meta-analysis. *J. Amer. Med. Assoc.* **277**: 739-745.

Sammartano L.J., Malejka-Giganti D., 1991, Interaction of C-nitroso aromatics with polyunsaturated fatty acids: Route to lipid peroxidation. *Chem-Biol. Interac.* **77**: 63-79.

Stasko, A., Brezova, V., Biskupic, S., Ondrias, K., Misik, V., 1994, Reactive radical intermediates formed from illuminated nifedipine. *Free Rad. Biol. Med.* **6**: 545-556.

Sullivan, A.B., 1966, Electron spin resonance studies of a stable arylnitroso-olefin adduct free radical. *J. Org. Chem.* **31**: 2811-2817.

Thomas, S.E., Wood, M.L., 1986, Photosensitivity reactions associated with nifedipine. *Brit. Med. J.* **292**: 992.

In vivo Detection of Nitric Oxide: Combining EPR and NMR

LAWRENCE J. BERLINER AND HIROTADA FUJII
Department of Chemistry, The Ohio State University, Columbus, Ohio 43210, USA and Sapporo Medical University, Sapporo, Hokkaido, Japan

1. INTRODUCTION

Nitric oxide (NO) is a biological regulator, originally called EDRF (endothelial derived relaxing factor), which is involved intimately in blood vessel relaxation and maintenance of blood pressure (Bredt and Snyder, 1989). NO is produced in the brain as a synaptic neuronal messenger (Palmer *et al.*, 1987). The enzyme nitric oxide synthase (NOS), which catalyzes the production of NO from arginine, is distributed constuitively in the vascular endothelium, brain, lung and platelets (cNOS), which also occurs as an inducible form, iNOS, when triggered by several factors such as cytokines and endotoxin, ie., lipopolysaccharide or LPS (Xie *et al.*, 1993).

Direct detection of the paramagnetic NO molecule by EPR (electron paramagnetic resonance) is possible in the gaseous phase; however dissolved NO is, effectively, EPR silent unless it can be "trapped" as an iron-dithiocarbamate complex utilizing the water soluble dithiocarbamates, MGD (N-methyl-D-glucamine dithiocarbamate), DTCS (succinodithiocarbamate) or the lipophilic DETC (diethyldithiocarbamate).

For example, Lai and Komarov (1994) observed NO production in septic shock rats as the $(MGD)_2Fe(II)$-NO complex in the tail blood circulation. Vanin and colleagues reported NO generation in tissue using $(DETC)_2Fe(II)$ as the complexing agent (Kubrina *et al.*, 1992). The aim of our research has been in vivo detection (or where necessary, excised tissue or ex vivo detection) by EPR spectroscopy.

Supramolecular Structure and Function 7, edited by Pifat-Mrzljak.
Kluwer Academic/Plenum Publishers, New York, 2001.

2. IN-VIVO EPR

The challenge of measuring paramagnetic signals in living animals has been successfully met over the past 15 years. Specifically, this was the development at low field rf (200 - 750MHz) and L-band spectrometers (1 - 2 GHz) with specialized large volume resonators and surface coils that could accommodate small animals (mice and rats). The frequency (wavelength) eventually limits the sample size, which equates to perhaps rabbits and rats, the results of which are yet to be realize. A complete treatise on the subject has recently been published (Berliner, 2000). Most of the work described here has been measured on homemade L-band EPR spectrometers operating at 1.2 – 1.24 GHz (Berliner and Koscielniak, 1991) as well as commercial X-band (9.5 GHz) spectrometers.

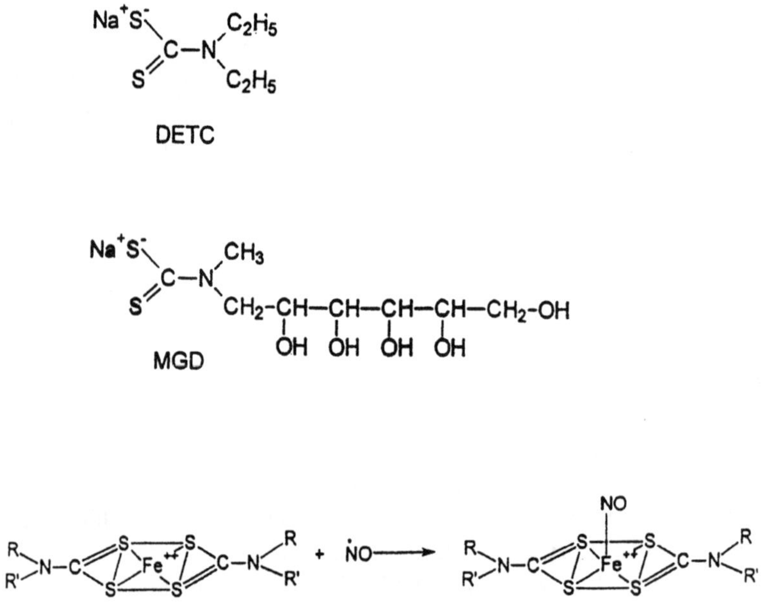

Figure 1. Structures of the dithiocarbamates and the Fe(II)-NO complex.

2.1 In-vivo detection of NO spin trap complexes

Figure 1 depicts the structures and complexes of the dithiocarbamate "spin trap" described above. In order to monitor biologically generated NO, it is necessary to induce high levels of NO's by LPS treatment. After six hours animals are under sepsis where high levels of iNOS (and consequently NO) are produced at high μM levels. Figure 2A depicts in vivo L-band spectra of a mouse localized in the upper abdominal region (near the liver) compared with an X-band spectrum of the excised liver. The higher sensitivity at X-band is obvious (Figure 2B); however, the line shapes at both L- and X-band are identical (Fujii *et al.*, 1999). The distribution shows that the highest levels are found in the liver. No trapped nitric oxide was detected in brain issue due to the fact that MGD or $(MGD)_2$ Fe(II) cannot cross the blood brain barrier.

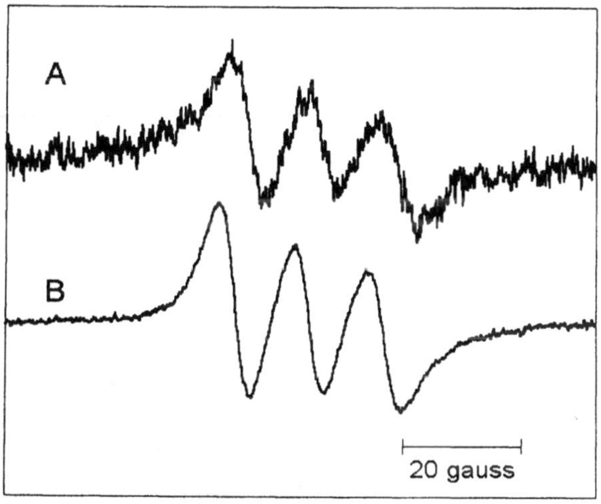

Figure 2. EPR spectra of $(MGD)_2$-Fe(II)-NO in LPS-treated mice. (A). In vivo L-band EPR spectrum of LPS-treated mice. Mice (about 20 g) were injected i.p. with LPS (1 mg in 0.3 ml of saline), followed 6 h later by subcutaneous injection with 0.4 ml of the NO spin-trap, $(MGD)_2$-Fe(II), in saline (MGD:100 mM, Fe:20 mM). The EPR spectrum was measured 2 h after administration of the NO spin-trap. (B) Ex-vivo X-band EPR spectrum of the excised liver from LPS-treated mice. Injection of LPS and NO spin-trap was carried out as noted above. One hour after spin trap injection, the mice were sacrificed under anesthesia, their organs excised and measured by X-band EPR. This spectrum represents a 12.5 mg liver sample. From Fujii *et al.* (1999) with permission.

On the other hand, DETC, which is actually the thiol reduced form of disulfuram (Antibuse), easily crosses the blood brain barrier as evidenced by its use with alcoholics to dissuade them from consuming intoxicating drinks. Figure 3 depicts X-band EPR spectra of excised organs and brain slices from animals (rats) which have been induced by LPS and subsequently injected with DETC and iron (II) citrate prior to EPR analysis. Note the relatively strong signal from the cerebellum as well as the heart and liver, where the NO was complexed within the lipid since $(DETC)_2Fe(II)$ is insoluble in

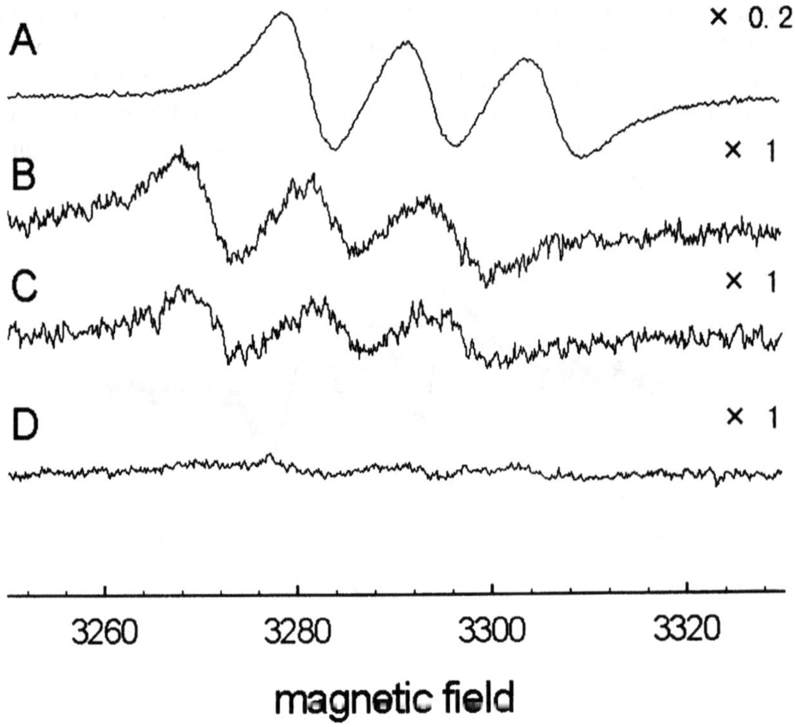

Figure 3. X-band EPR spectra of $(DETC)_2$-Fe(II)-NO in excised tissues after in vivo administration of LPS, DETC and ferrous citrate to rats. EPR spectra of $(DETC)_2$-Fe(II)-NO in (A) liver (receiver gain; 5 x reduction), (B) hippocampus, (C) cerebellum, and (D) olfactory bulb. Rats were injected i.p. with LPS (50 mg/kg) in saline, and after 6 h DETC (500mg/kg) was injected i.p. and then a mixture of ferrous sulfate (100 mg/kg) and sodium citrate (500 mg/kg) was injected subcutaneously into rats. After 1 h following the injection, rats were sacrificed under anesthesia, and each tissue was immediately excised and kept on dry ice until measurements of EPR spectra. Controls without LPS injection showed very low intensity EPR spectra. Typical spectrometer conditions were: microwave power, 20 mW; 100 KHz modulation amplitude, 1.0 gauss; sweep rate, 50 gauss/min; response time, 0.3 s.

aqueous regions. (In fact, the solubility of both NO and O_2 are relatively higher in lipid vs. polar environments). The second point to note is the effect of the specific NOS inhibitor, L-monomethylarginine (LMMA) on suppressing NO production (Figure 4).

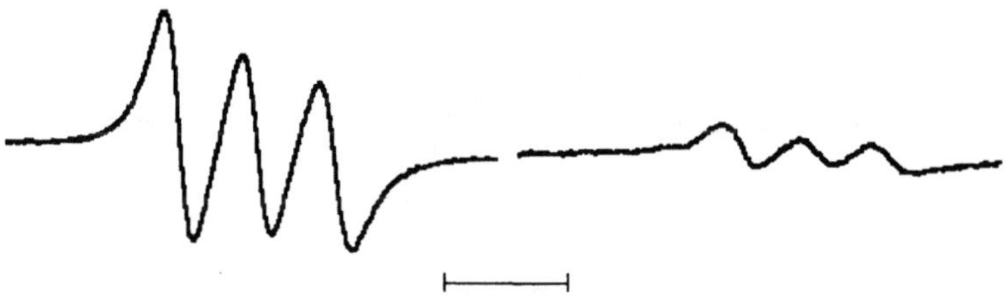

Figure 4. Inhibitory effect of L-NMMA on the NO generation in LPS-treated rats. X-band EPR spectra of NO complex were measured in excised liver tissue from LPS-injected rats. Injection of LPS, DETC and ferrous citrate was carried out as denoted in Figure 3 in the absence (left spectrum) and presence (right spectrum) of L-NMMA (50mg/kg in saline), administered i.p. 3 h after LPS injection.

2.2 Suppression of iNOS expression

The spin trap, phenyl-N-*tert*-butylnitrone (PBN) was found to block iNOS expression at the sign along or genetic level presumably via the NfκB pathway (Miyajima and Kotake, 1995). In order to determine whether this phenomenon occurs in whole animals, we pretreated mice with PBN prior to LPS injection as well as control experiments where either no PBN was administered or was administered 2 hours after LPS induction ((Fujii *et al.*, 1997). As shown in Figure 5, total suppression of NO production was observed when animals were pretreated with PBN, while post-LPS injection was ineffective since the signaling pathway had already been triggered. In order to confirm that the NO produced was indeed from NOS and not another nonenzymatic pathway such as the reduction of nitrite, animals were treated with ^{15}N-L-arginine, which would yield ^{15}NO in the iron-dithiocarbamate complex. Figure 5d depicts the resultant EPR spectrum which is a summation of both two-line ^{15}NO and three-line highly abundant ^{14}NO EPR spectral components. Figure 6 depicts analogous experiments on brain tissue where the DETC spin trap was employed (Fujii and Berliner, 1999). Remarkably, NOS expression was not suppressed in brain vs. other tissue! There are several possible explanations for this surprising result: either the signaling pathway for iNOS expression is totally different in the brain, or even though PBN passes the blood brain barrier it was inaccessible to the iNOS signaling pathway in *any* substructure of the brain. It is remarkable nonetheless, that this difference was discovered from in-vivo/ex-vivo EPR studies.

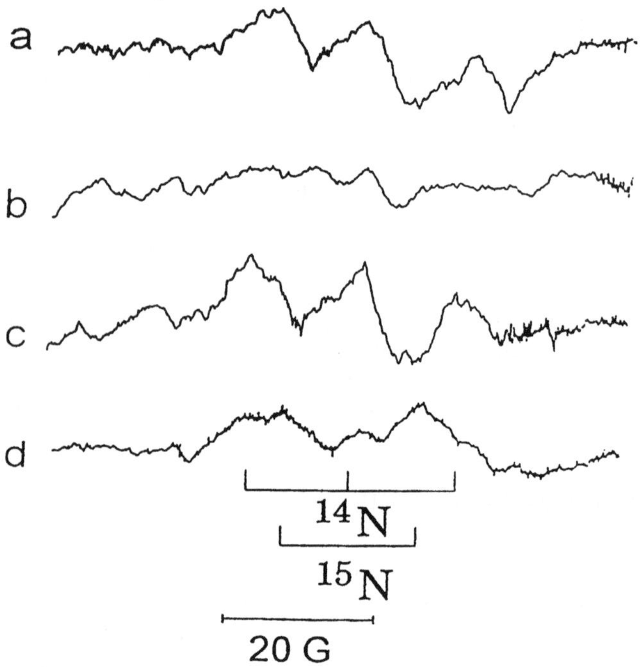

Figure 5. In vivo L-band EPR spectra of $(MGD)_2$-Fe(II)-NO in LPS-treated mice. (a) A mouse (ca. 20g) was injected i.p. with LPS in saline (1 mg/0.3ml), and after 6hrs a subcutaneous injection with 0.4 ml of $(MGD)_2$-Fe(II) complex in saline [MGD:Fe(II), 100 mM:20mM]. After 2 hrs following the last injection the EPR spectrum of the mouse was measured. (b) Same as (a), except that PBN (6mg/0.5 ml saline) was injected i.p. into the mouse 30 min *before* LPS administration. (c) Same as (a), except that the PBN was injected i.p. 2 hrs *after* LPS administration. (d) Same as (a), except that ^{15}N-L-arginine (10 mg/0.3 ml saline) was injected i.p. into the mouse 6 hrs after LPS administration, and just prior to the injection of the $(MGD)_2$-Fe(II) complex. Spectrometer conditions were: frequency, 1.256 GHz; applied magnetic field, 425 gauss; incident microwave power, 100 mW; modulation, 0.5 gauss; sweep rate, 50 gauss/min.

3. VISUALIZING NO BY MAGNETIC RESONANCE IMAGING

An important question is precisely where NO generation (and accumulation) occurs as well as more detailed pathology at to the substructural level. In fact, Yoshimura and colleagues reported EPR images

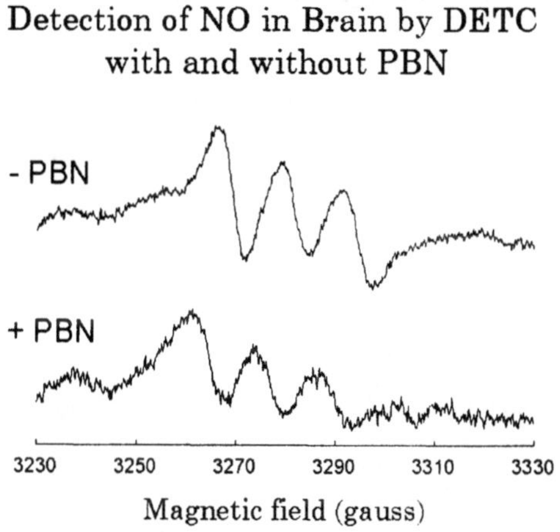

Figure 6. Effects of PBN on NO generation in LPS-treated rats. X-band EPR spectra were measured in tissues excised from LPS-injected rats. Rats were injected with LPS , DETC and ferrous citrate. In the PBN experiments, the nitrone (250 mg/kg) was pre-administered i.p. in rats 30 min before LPS injection. Saline solutions containing no PBN were injected as a control. EPR spectra are shown from rat cerebellum.

of $(DTCS)_2$ Fe(II)-NO in the upper abdominal region of a septic shock mouse (Yoshimura *et al.*, 1996). However, the spatial resolution of in vivo EPR spectrometers is not high enough to resolve detailed fine structure of specific organs visualizing paramagnetic NO complex distribution, especially since the intrinsic line width of e.g. NO complex is quite large. On the other hand, MRI instruments can produce image resolution that compares with CAT scanning. Furthermore, paramagnetic contrast agents enhance image resolution by proton relaxation enhancement of surround water protons, as shown with nitroxyl spin probes (Bennett *et al.*, 1990; Chen *et al.*, 1990; Slane *et al.*, 1986). The $(MGD)_2$ Fe(II) NO complex is a good relaxant. The strong magnetic moment of the unpaired electron promotes shortening of the spin-lattice (T_1) and spin-spin (T_2) relaxation times at both 20 and 85 MHz. The T_1 relaxivity was 0.31 and 0.27(1/mM-sec) respectively while the T_2 relaxivity was 0.31 and 0.35 (1/mM-sec) respectively at the same two frequencies, respectively. Hence the relaxivity is not field dependent in this range (Fujii *et al.*, 1999). These relaxivities were similar to those found with nitroxyl compounds (Bennett *et al.*, 1990; Chen *et al.*, 1990; Slane *et al.*, 1986), but have the distinct advantage of higher resistance to biological reduction (or oxidation) in vivo over the periods required for imaging. Furthermore, this relaxation enhancement occurs only for the NO complex; the T_1 relaxivity of $(MGD)_2Fe(II)$ alone was 0.044 (1/mM-sec) at both 20 and 85 MHz, which is essentially negligible.

These properties suggest that $(MGD)_2Fe(II)$-NO should be a unique, effective contrast agent for MRI. Figure 7 depicts T_1-weighed MR images of the axial plane of the liver of Wistar rats which have been injected with

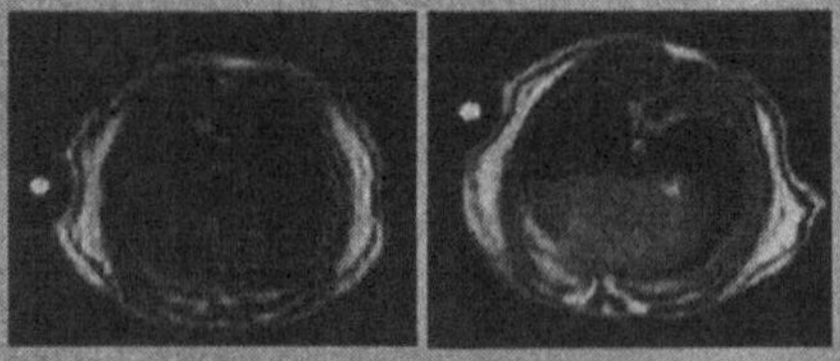

Figure 7. T_1-weighted MR images in the axial plane of the liver of Wistar rats. (A) Control; (B) 60min after injection of preformed $(MGD)_2$-Fe(II)-NO complex. Two ml of 9.5 mM $(MGD)_2$-Fe(II)-NO complex, made from NO gas and $(MGD)_2$-Fe(II), was injected i.p. at the lower abdomen of rats (250 g). From Fujii *et al.* (1999) with permission.

preformed $(MGD)_2Fe(II)$-NO complex. The image on the right clearly shows that this NO complex is an effective, NO specific, "intrinsic contrast agent," resulting in enhanced contrast in several organs. The concentration of NO complex was around 150 ± 52nmol/g of tissue from EPR assays of excised liver (Fujii *et al.*, 1997; Fujii *et al.*, 1999). In order to test whether biologically generated NO could be "visualized" by MRI we injected rats with LPS and then measured by MRI using a protocol analogous to that used for EPR. Figure 8 shows the resultant MR images. Again, the liver was clearly distinct from other organs. The image intensity reached a maximum at about 90 - 120 min after injection of $(MGD)_2Fe(II)$ for the particular slice shown. In order to confirm that NO generation occurred at this site *vs.* an accumulation of $(MGD)_2Fe(II)$-NO in the liver from the systemic circulation,

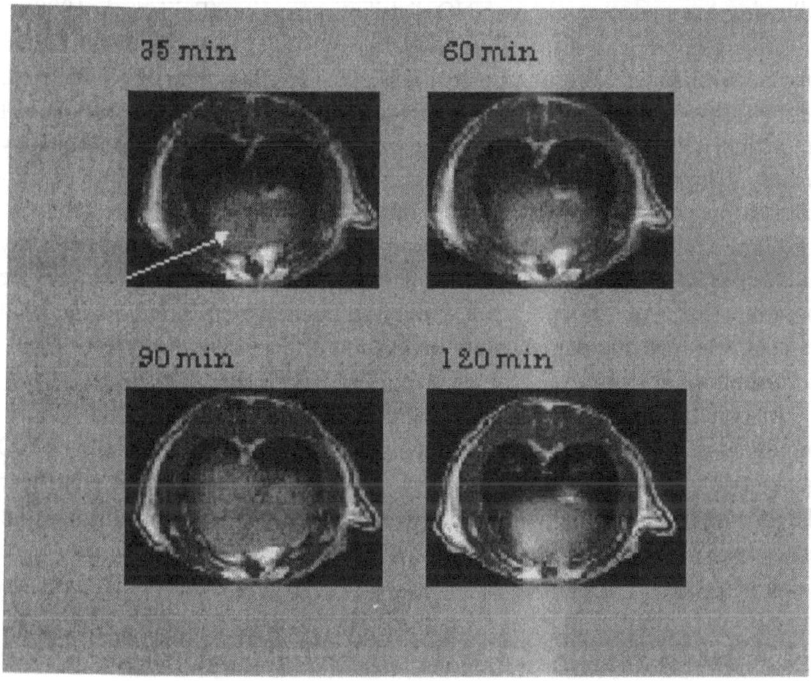

Figure 8. Transverse T_1-weighted MR images of the liver in LPS-doped rats. MR images were measured at the times indicated for two different slices of the liver. Six hours after LPS injection, the NO spin-trap (3 ml of $(MGD)_2$-Fe(II), MGD: 100 mM, Fe: 20 mM) was administered i.p. The slice thickness was 2 mm, and each slice was separated by 1 mm. The image enhancement was drastically reduced upon treatment with the NOS inhibitor, L-NMMA.

we simultaneously monitored NO levels by EPR at both the liver and the tailblood circulation. The results confirmed that NO generation in the liver maximized 90-120 min. after spin trap injection while the systemic levels did not plateau until 150-180 min (Fujii *et al.*, 1997). One can also follow the biochemistry of liver iNOS by MRI by administering a specific NOS inhibitor, LMMA, where suppression of image contrast was clearly evident when LMMA was administered, verifying that the NO production was from NOS and not produced via other mechanisms, such as nitrite reduction.

3.1 MRI spin trapping

The experiments depicted above demonstrate the feasibility of using MRI strategies combined with spin trapping in order to map "NO distributions" in tissues and organs at higher spatial resolution. In fact the "spin trapped adduct" $(MGD)_2$ Fe(II)-NO functioned as a specific NMR contrast agent, allowing one to accurately map NO distribution in vivo (Fujii *et al.*, 1999). In principle, one can also follow the kinetics of NO generation at different organs with superior resolution than currently possible with EPR. We have termed this technique as "MRI spin trapping," a subset of the general approach, 'NMR spin trapping,' which is discussed briefly in the work by Valery Khramtsov which encompasses the use of ^{31}P- and ^{19}F-NMR of phosphorus or fluorine containing nitroxyl spin traps (Khramtsov *et al.*, 1999). Functional MRI (f MRI) measures changes in magnetic susceptibility as a result of changes in blood flow of oxygenated hemoglobin. It may be possible that spin trapped NO, which reflects blood flow independent effects, may provide paramagnetic relaxation effects as a new type of f MRI contribution.

In conclusion, this work demonstrates the power of localized EPR in monitoring NO production in vivo as well as mapping and visualizing NO generation in septic shock rats by MRI. The feasibility has been demonstrated; it only remains for new experimentation to expand the scope and power of this marriage of NMR and EPR.

REFERENCES

Bennett, H.F., Brown, R.D. III, Keana, J.F., Koenig, S.H., and Swartz, H.M., 1990, Interactions of nitroxides with plasma and blood: effect on $1/T_1$ of water protons. *Magn. Reson. Med.* **14**: 40-55.

Berliner, L.J., and Koscielniak, J., 1991, Low field EPR spectrometers: L-band. in *"EPR Imaging and In-Vivo EPR"* (G. R. Eaton, S. S. Eaton, K. Ohno, Eds.) CRC Press, Inc., Boca Raton, pp. 65-72.

Berliner, L.J. (ed), 2000, *In Vivo EPR: Theory and Applications,* Kluwer-Plenum Academic Publishing, New York..

Bredt, D.S., and Snyder, S.H., 1989, Nitric oxide mediates glutamate-linked enhancement of cGMP levels in the cerebellum. *Proc. Natl. Acad. Sci. USA* **86**: 9030-9033

Chen, H.C., Sun, K.Q., Magin, R.L., and Swartz, H.M., 1990, Potential of albumin labeled with nitroxides as a contrast agent for magnetic resonance imaging and spectroscopy. *Bioconjugate Chemistry* **1**: 32-36.

Fujii, H., and Berliner, L.J., 1999, Ex-vivo detection of nitric oxide in brain tissue. *Magnetic Resonance Medicine* **42**: 599-602.

Fujii, H., Koscielniak, J., and Berliner, L.J., 1997, Determination and characterization of nitric oxide generation in mice by in-vivo L-band EPR spectroscopy. *Magnetic Resonance Medicine* **38**: 565-568.

Fujii, H., Wan, X., Zhong, J., Yoshikawa, K., and Berliner, L.J., 1999, In vivo imaging of spin-trapped nitric oxide in rats with septic shock: MRI spin trapping. *Magnetic Resonance in Medicine* **42**: 235-239.

Khramtsov, V.V., Berliner, L.J., Clanton. T.L., 1999, NMR spin trapping of free radical reactions using a phosphorus-containing nitrone spin trap. *Magn. Reson. Med.* **42**: 228-234.

Kubrina, L.N., Caldwell, W.S., Mordvintcev, P.I., Malenkova, I.V., and Vanin, A.F., 1992, EPR evidence for nitric oxide production from guanidino nitrogens of L-arginine in animal tissues in vivo. *Biochim. Biophys. Acta* **1099**: 233-237.

Lai, C-S., and Komarov, A.M., 1994, Spin trapping of nitric oxide produced in vivo in septic-shock mice. *FEBS Lett.* **345**: 120-124.

Miyajima, T., and Kotake, Y., 1995, Spin trapping agent, phenyl N tert-butyl nitrone, inhibits induction of nitric oxide synthase in endotoxin-induced shock in mice. *Biochem. Biophys. Res. Comm.* **215**: 114-121.

Palmer, R.M., Ferrige, A.G., and Moncada, S., 1987, Nitric oxide release accounts for the biological activity of endothelium-derived relaxing factor. *Nature* **327**: 524-526.

Slane, J.M., Lai, C.S., and Hyde, J.S., 1986, A proton relaxation enhancement investigation of the binding of fatty acid spin labels to human serum albumin. *Magn. Reson. Med.* **3**: 699-706.

Xie, Q.W., Whisnant, R., and Nathan, C.F., 1993, Promoter of the mouse gene encoding calcium-independent nitric oxide synthase confers inducibility by interferon gamma and bacterial lipopolysaccharide. *J. Exp. Med.* **177**: 1779-1784.

Yoshimura, T., Yokoyama, H., Fujii, S. ,Takeyama, F., Oikaw, K., and Kamada, H., 1996, In vivo EPR detection and imaging of endogenous nitric oxide in lipopolysaccaride-treated mice. *Nature Biotech.* **14**: 992-994.

On Biological Information*

MANFRED EIGEN
Max-Planck-Institute for Biophysical Chemistry, Göttingen/Germany

What is the distinguishing feature of a living system that singularizes it from every non-living chemical ensemble, regardless of the extent of its complexity. The differentiable characteristic of the living system is information. Information assures the controlled reproduction of all the constituents, thereby ensuring the conservation of viability. Information - unlike energy - is not subject to a conservation law. Hence the fundamental question behind the origin of life is: How can information originate? Information theory, which was pioneered by Claude Shannon, cannot answer this question. Shannon's theory is most successful in dealing with problems of coding and transmission of information.

Information has two aspects: A quantity to be called "extent" and a quality we call "content". Extent appears as an entropy measure in Shannon's theory. Content, which deals with meaning, originates via selective self-organization. It can be described also in quantitative physical terms. A prerequisite is the reproducibility of the informational substrate forming the basis of selection. The information that is unique for life evolves by virtue of "natural selection". Today we can be more specific: Natural selection is a non-equilibrium process and can be considered an analogue of a physical phase transition. As such it is an inherent consequence of

* This extended abstract summarizes the three lectures given:
1.) "The origin of biological information"
2.) "Evolutionary biotechnology"
3.) "Molecular diagnostics: Viruses, prions and Alzheimer's disease."

Supramolecular Structure and Function 7, edited by Pifat-Mrzljak.
Kluwer Academic/Plenum Publishers, New York, 2001.

mutagenous self-replication at several levels of organization: For instance, it is evident in molecules such as nucleic acids, in molecular complexes such as viruses and in autonomous forms of life such as micro- or higher organisms. New physical concepts have been introduced in order to deal quantitatively with the dynamics of the molecular generation of genetic information. They provide a physical foundation for Darwinian behaviour, yet they introduce major modifications in its interpretation. The lecture deals with these physical concepts, such as "sequence space" and "quasi-species".

Complexity of biological information is determined by the length of the coding molecules (RNA or DNA) and their composition. In the case of nucleic acids, the number of all possible combinations is 4^{ν}. To arrange the resulting enormous number of different possible sequences, the concept of sequence space was introduced: Each sequence occupies one single point in this space, the distance between them is a measure of their homology.

As a consequence, the sequence space is characterized by (relatively short) mutual distances between any two points, a huge storage capacity and a tremendous connectivity. The quasispecies, a mutant clan having a defined master sequence, can be described as a "cloud" in this space wandering to higher fitness levels. We call this "cloud" quasispecies because it can be described mathematically by one normal mode which behaves like one defined species.

Experimental evidence for the quasispecies has been obtained for RNA-viruses in particular for the bacterio-phage Qß. In their hosts these viruses are present as huge mutant distributions generated by an error-prone replication of the genes that encode the vital enzymes (error rate of Qß-replicase $\approx 3 \times 10^{-4}$). Data have shown that Qß works exactly below its error-threshold, a critical value above which replication would lead to nonsense products (equivalent with a "melting" of the information contained in the consensus sequence of the mutant distribution).

The best-adapted sequence (master sequence) usually constitutes only a minor fraction of the virus quasispecies distribution. Its fraction strongly depends on the error rate of the replicating enzyme: With increasing error rate it is replaced by mutants and near the error-threshold its fraction becomes quite small. All viruses show mutant distributions in good agreement with the quasispecies model.

Is it possible to identify single molecules of a quasispecies distribution?

Fluorescence correlation spectroscopy offers the opportunity to detect single molecules if they are labeled with a fluorescent dye. In this technique a laser-beam is directed into the sample where it generates a focus in a volume element containing a fraction of a femtoliter. The molecules, which diffuse through this volume element are excited and their fluorescence intensity is detected in a time resolved manner. The autocorrelation of these intensity signals generate a correlation curve, which can be evaluated mathematically yielding the diffusion time and the fraction of the fluorescent molecules present in the solution.

Combined with amplification by PCR polymerase chain reaction or NASBA (nucleic acid sequence-based amplification) FCS enables the detection of RNA-/DNA-template species down to the level of single molecules: A third fluorescently labeled probe is integrated in the amplification reaction. Its extension during this reaction is monitored *online* by FCS. On the basis of this technique, a new highly sensitive test for HIV-1 in blood plasma was developed which extends down to few virus particles per ml. A molecular diagnostics of virus diseases has become reality.

Another application of "molecular diagnostics" deals with prions. The diagnosis of prion diseases (such as Creutzfeldt-Jakob disease (CJD) in man) relies on the detection of the pathological prion protein (PrP^{Sc}). This protein is present in the cerebrospinal fluid. However, no quantitative test for PrP^{Sc} was available thus far. Based on confocal dual-color fluorescence correlation spectroscopy, such a highly sensitive method for quantitative detection of PrP^{Sc} in spinal liquor has been developed in our laboratory. Pathological prion protein aggregates were labeled by two specific antibodies which were tagged with two different fluorescent dyes, yielding intensely fluorescent targets. These aggregates can be detected by their dual-color fluorescence intensity. Analysis of the distribution of fluorescence signal coincidences allows for a sensitive identification of such pathological aggregates. In a diagnostic model system, PrP^{Sc} aggregates were detected down to a concentration of 2 pM PrP^{Sc}, corresponding to an aggregate concentration of approximately 2 fM. The method exceeds all existing procedures in its sensitivity of detection. The PrP^{Sc}-specific signals could only be detected in samples from patients with CJD, but not in control samples. Furthermore, this method could also be adapted to a sensitive detection of other disease-associated amyloid aggregates such as found in Alzheimer's disease.

The method of "molecular diagnostics" has also important applications in biotechnology. It provides the basis for "high throughput screening" in drug discovery and as such is instrumental in finding optimally adapted structures in molecular evolution.

REFERENCES

Bieschke, J., Giese, A., Schulz-Schaeffer, W., Zerr, I., Poser, S., Eigen, M. and Kretzschmar, H., 2000, Ultrasensitive detection of pathological protein aggregates by dual-color scanning for intensely fluorescent targets, *Proc. Natl. Acad. Sci.* USA **97**: **5468-5473**.

Brakmann, S. and Eigen, M., 1996, Evolution in the test tube, *Frontiers in Biology*, Vol. 1 (eds. W. Gilbert and G. Tocchini, Valentini), pp. 243-257.

Eigen, M. and Gardiner, W.C., 1984, Evolutionary molecular engineering based on RNA replication, *Pure Appl. Chem.* **56**: 967-968.

Eigen, M. and Nieselt-Struwe, K., 1991, How old is the immunodificiency virus? *AIDS* **5**: 585-593.

Eigen, M. and Rigler, R., 1994, Sorting single molecules: Application to diagnostics and evolutionary biotechnology. *Proc. Natl. Acad. Sci.* USA **91**: 5740-5747.

Eigen, M., 1993, The origin of genetic information: Viruses as models. *Gene* **135**: 37-47.

Eigen, M., Biebricher, C.K., Gebinoga, M. and Gardiner, W.C., 1991, The hypercycle - coupling of RNA and protein biosynthesis in the infection cycle of an RNA-bacteriophage. *Biochemistry* **30**: 11005-11018.

Eigen, M., McCaskill, J. and Schuster, P., 1988, Molecular quasispecies. *J. Phys. Chem.* **92**: 6881-6891.

Eigen, M., McCaskill, J. and Schuster, P., 1989, The molecular quasi-species. *Adv. Chem. Phys.* **75**: 149-263.

Koltermann, A., Kettling, U., Bieschke, J., Winkler, T., and Eigen, M., 1998, Rapid assay processing by integration of dual-color fluorescence cross-correlation spectroscopy: High throughput screening for enzyme activity. *Proc. Natl. Acad. Sci.* USA **95**: 1421-1426.

Leuthäusser, I.,. 1986, An exact correspondence between Eigen's evolution model and a two-dimensional Ising system. *Chem. Phys.* **84**: 1884-1885.

Oehlenschläger, F., Schwille, P. and Eigen, M., 1996, Detection of HIV-1 RNA by nucleic acid sequence amplification combined with fluorescence correlation spectroscopy. *Proc. Natl. Acad. Sci.* USA **93**: 12811-12816.

Accuracy *versus* Efficacy in Biological Processes*

MIROSLAV RADMAN
INSERM E9916, Faculté de Médecine Necker-Enfants Malades, Université de Paris-5, 156 rue de Vaugirard, 75015 Paris, France

1. INTRODUCTION

The complexities of the cellular biochemistry and of the architecture of living creatures were traditionally seen as the demonstration of a supreme accuracy in the expression of genetic information. Genetic information itself was considered as being close to perfect – the result of billions of years of evolution - and therefore tending always towards minimal mutation rates. However, in real life, survival rather than perfect fidelity is the ultimate virtue, and survival results from robustness and adaptability to an unpredictable environmental future. Because adaptability involves exploration, molecular infidelity coupled with repetition often makes success more certain than would a very precise non-repetitive process (Radman, 1999).

There is ample evidence that the accuracy of life processes is not maximised but optimised because of the trade-off between accuracy and efficacy. In fact, life structures do not arise by a fully deterministic genetic design. The precise molecular event, or structure, is typically the result of a process that is often a succession of trials and errors. Embryonic development, and the frequency of its failures (at least 50% in humans), is a *par excellence* example of life's imperfections (Vogel and Motulsky, 1996). Another example is the process of chromosome segregation from a mother cell into its two daughter cells. The polymerising fibres (microtubules) do

* This paper is dedicated to the memory of my first science teacher, who became a precious mentor and friend, Dr. Marija Badel-Drakulic.

Supramolecular Structure and Function 7, edited by Pifat-Mrzljak.
Kluwer Academic/Plenum Publishers, New York, 2001.

not "know" where exactly is their chromosomal target (centromere) to shoot for: they keep shooting and missing until the winning hit. A very precise one-time shot would often lead to a "precise missing" of a target with an uncertain position, whereas successive imprecise missings will sooner or later lead to the hit.

2. FIDELITY OF BIOSYNTHETIC PROCESSES

How about the fidelity of enzymatic reactions? John Hopfield (1974) and Jacques Ninio (1975) provided the concept of "kinetic proof-reading" in biochemical reactions: cellular enzymes make mistakes all the time but each mistake (e.g., a complex between the enzyme and a wrong substrate) slows down the next reaction step and is therefore either abandoned, rectified, or actively destroyed. The existence of error-destruction and error-correction mechanisms – the molecular repair systems - allows for the high speed of biochemical reactions, i.e., for their efficacy, without fatal accumulation of errors. An excessively precise process would be prohibitively slow and/or energy consuming, i.e., inefficient (Ninio, 1997). In reality, all fundamental biological processes, including the replication of genetic information, proceed efficiently - and therefore relatively imprecisely – leaving the errors to the specialised error-correction enzymes that hydrolyse the wrong residue or the whole intermediate (see Figure 1, Friedberg *et al.*, 1995: Radman *et al.*, 2001). Such enzymes are themselves quite efficient because their substrate is specifically the mistake in some other enzyme's reaction.

What are the error rates in the biosynthesis of informational macromolecules in living cells? For protein biosynthesis about 3 erroneous amino acids per 10 thousand correct ones; for transcription about 1 wrong nucleotide per 100 thousand, and for DNA replication about 5 wrong nucleotides per 10 billions! Proteins are the functional molecules that get used up, oxidised, broken down and replenished by *de novo* synthesis. As many as 30% of all proteins synthesised by normal human cells are immediately degraded by proteasomes because they carry recognisable folding mistakes leading to their tagging by ubiquitination (Schubert *et al.*, 2000). Up to 50% of all mRNAs are erroneous and therefore broken down by RNases (Alberts *et al.*, 1994). In contrast, there is clearly a major investment in the DNA fidelity because genes are the memory of evolution, Life's database for protein structure. And yet, even the DNA fidelity is maintained by the enzyme activities (see Fig 1 and below).

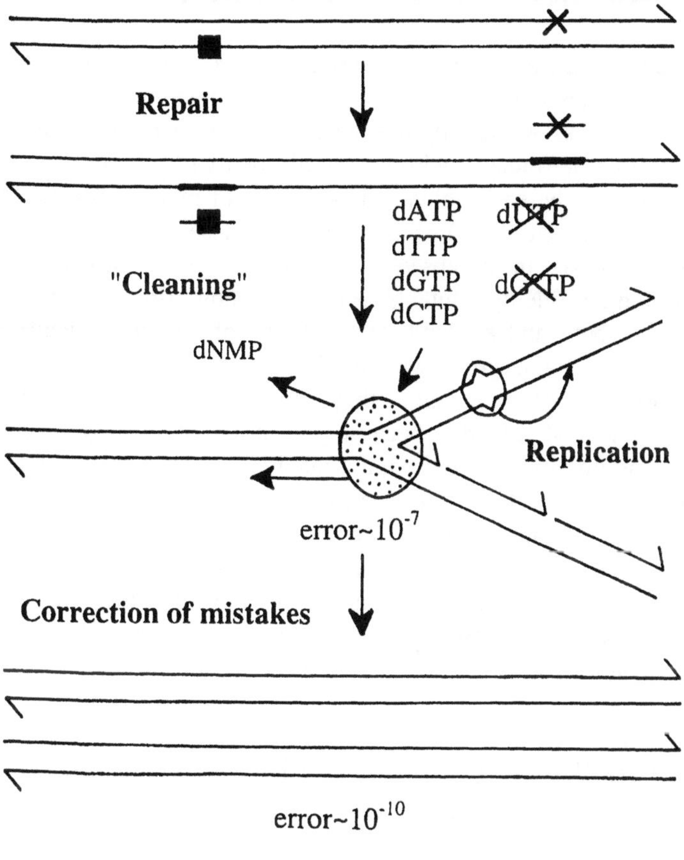

Figure 1. Error avoidance and error correction in DNA replication. The basic strategies to conserve the DNA sequence involve: (a) The maintenance of the chemical purity of ingredients for DNA replication, i.e., the repair of DNA templates and the enzymatic chemical purification and equilibration of the four substrates for making DNA (dNTP pools) ; (b) The high fidelity DNA copy machine, i.e., efficient selection of correct dNTPs by DNA polymerase and the immediate removal of the emerging mistake by the proofreading exonuclease and (c) The quality control of new strands, i.e., the postreplicative mismatch repair of the new strands rectifying copy errors according to the original template strand Radman *et al.*, 1980). The latter system lowers the frequency of spontaneous mutations by at least 100-fold (Glickman and Radman, 1980). See Friedberg *et al.* (1995) for more references.

3. ACCURACY *VERSUS* EFFICACY

The question arises whether the efficacy (and precision!) of the error correction enzymes is maximised, i.e., is nature attempting perfect fidelity via repair enzymes? Apparently not, because it is possible to obtain, in the laboratory, bacterial mutants with increased fidelity of protein biosynthesis. Such mutants can be easily obtained as a class of mutants resistant to the antibiotic streptomycin (StrR) which at sublethal doses increases error rate in protein biosynthesis. In the absence of streptomycin, the StrR mutants are (ribosomal) high fidelity mutants which all show slowed-down growth (review, Ehrenberg and Kurland, 1984). Some of the strR mutants are so slow that they cannot grow without the presence of an error-inducing concentration of streptomycin: they are streptomycin-dependent or "addicted" mutants! Clearly, because the price of fidelity is paid by efficiency, the fidelity is optimised, not maximised. Although more difficult to obtain, bacterial mutants with increased DNA polymerase fidelity have been described. Of course, too high error rates would ruin efficacy and even survival: it was shown that bacteria and haploid yeast die from genetic catastrophe when their mutation rate is increased by 104-fold. The molecular error catastrophe theory, based on feedback effects of increased error rates in protein biosynthesis, was developed by L. Orgel (1963, 1970).

4. FIDELITY OF DNA REPLICATION AND THE STORY OF SOS POLYMERASES

Optimised fidelity is a general principle that applies also to genes. When adaptation is limited by the available genetic diversity, genetically unstable bacterial populations are expected to adapt, at least short term, more rapidly than their stable counterparts (Taddei *et al.*, 1997; Giraud *et al.*, 2001). Two classes of genes are known to accelerate genetic variation by enhanced mutation and/or recombination rates in bacterial populations: the stress-inducible wild type genes, usually part of the SOS regulon, and genes whose functional loss, or down regulation, increases the rate of genetic variability (mutator and/or hyper-rec mutants). These two evolution-driving strategies seem to coexist in natural populations of bacteria, one acting at the level of individual cells, the other at the population level (Radman *et al.*, 2000; 2001).

However, in addition to evolution, there is a down-to-earth reason for a certain level of DNA infidelity: the DNA replication itself! A top fidelity DNA copy machine, such as the DNA replication complex, can copy only a chemically perfect DNA, which certainly is not the case of cellular DNA. Intrinsic chemical instability of DNA is due to spontaneous oxidation, hydrolysis, alkylation, strand breaks etc. Estimates go up to 300.000 such lesions per human cell per day (review, Lindahl and Wood, 1999), all subject to special repair systems, but each unrepaired lesion can stop the regular DNA replication machinery – a potentially lethal event. The solution to this problem seems to be provided by a special class of unusual and fascinating DNA polymerases, SOS polymerases, which allow the trans-lesion DNA synthesis (TLS) assuring the completion of DNA replication in spite of the presence of DNA lesions (review, Hanaoka, 2001). If there are three SOS inducible DNA polymerases in *Escherichia coli*, the human genome sequences show at least 9 recognisable unusual DNA polymerases (in addition to 4 standard ones) (Wood *et al.*, 2001).

Three decades ago, the study of bacterial induced mutagenesis defined the TLS as an error-prone DNA synthesis - a key component of the stress-inducible cellular SOS response in bacteria that controls also the bacterial cell division (the original discovery of cell cycle control) (review, Radman *et al.*, 2001). Whereas inducible SOS polymerases are responsible for all induced mutagenesis in bacteria, their numerous eucaryotic homologues seem to be expressed all the time, presumably because large genomes always contain some unrepaired lesion and therefore require the activity of such polymerases for each S-phase. A delicate evolutionary question arises: did these SOS polymerases evolve to provide increased mutation rates under stress or to assure survival of cells with damaged DNA (see, Radman, 1999)? Recent discoveries brought new and unexpected insights (reviews, Friedberg *et al.*, 2000, Hanaoka, 2001). Although responsible for damage-induced mutagenesis, SOS polymerases seem to have evolved to copy damaged DNA templates by preventing what appears as unavoidable mutations. Here is the gist of the current thinking about this apparent paradox reviews. Each SOS polymerase appears to be specialised for a particular lesion because it can copy such lesion with a base that restores the original sequence: they are slow acting antimutagenic DNA polymerases, but only for their "cognate" lesion. However, being highly lesion-specialised, SOS polymerases are mutagenic when copying normal DNA or a non-cognate lesion (review, Radman *et al.*, 2001). Thus, although SOS polymerases have evolved most likely to allow replication of damaged DNA in the least mutagenic fashion, they are now required for all induced mutagenesis!. SOS polymerases provide an extraordinary example of the duality of fidelity and infidelity, stability and instability.

5. GENETIC INFIDELITY

Finally, even the genetic fidelity to one's own species is far from maximised, at least in bacteria. The mismatch repair system (Fig 1) is the functional genetic barrier between bacterial species (Matic *et al.*, 1995) and this is true also for meiotic crosses in yeast (Hunter *et al.*, 1996). Whereas the barrier to gene exchange between closely related species is removed in mismatch repair mutants (Rayssiguier *et al.*, 1989), an overproduction of the key mismatch repair proteins, MutS and MutL, creates strong genetic barriers between strains of the same bacterial species (Vulic *et al.*, 1997). Clearly, there was no selective pressure for "pure race" in bacteria: it would have been so easy to produce some more MutL/S to increase genetic fidelity! The unequalled evolutionary success of bacteria can be explained by their large population sizes and by their genetic infidelity: bacteria share an immense gene pool belonging to thousands of different species via their sexual promiscuity called horizontal gene transfer (Denamur *et al.*, 2000 and references therein).

Errors and infidelity, even wastefulness (viz. the immune system generating hundreds of millions mostly useless antibody genes), can cause individual failures but they are also the source of innovation and robustness assuring the perpetuation of life. Clearly, Mother Nature does not exhaust itself for the sake of fidelity and perfectionism. Rather, errors are made, often repaired or discarded, but always permanently tested as the source of blind innovation during the ongoing adaptation to unpredictable environmental changes and challenges.

ACKNOWLEDGMENTS

I indebted to my colleagues Ivan Matic and François Taddei who are for me a constant source of inspiration and challenge. Discussions with J. Ninio and E.C. Friedberg have influenced this review. My work is sponsored by INSERM, ARC-Villejuif and Necker foundations.

REFERENCES

Alberts, B., Bray, D., Lewis, J., Raff, M., Roberts, K. and Watson, J.D., 1994, *Molecular Biology of the Cell*, Garland, New York.

Denamur, E., Lecointre, G., Darlu, P., Acquaviva, C., Sayada, C., Sunjevaric, I., Rothstein, R., Elion, J., Taddei, F., Radman, M. and Matic, I., 2000, Evolutionary implications of the frequent horizontal transfer of mismatch repair genes. *Cell* **103**: 711-721.

Ehrenberg, M. and Kurland, C.G., 1984, Costs of accuracy determined by a maximal growth rate constraint. *Q. Rev. Biophys.* **17**: 45–82.

Friedberg, E.C., Walker, G.C. and Seide, W., *DNA repair and mutagenesis.* ASM Press, Washington.

Friedberg, E.C., Feaver, W.J. and Gerlach, V.L., 2000, The many faces of DNA polymerases: Strategies for mutagenesis and for mutation avoidance. *Proc. Natl. Acad. Sci. USA* **97**: 5681-5683.

Giraud, A., Matic, I., Tenaillon, O., Clara, A., Radman., M., Fons, M. and Faddei, F., 2001, Costs and Benefits of High Mutation Rates: Adaptive evolution of bacteria in the mouse gut. *Science* 2001, in press.

Glickman, B.W. and Radman, M., 1980, E. coli mutator mutants deficient in methylation instructed DNA mismatch correction. *Proc. Natl. Acad. Sci. USA* **77**: 1063-1067.

Hanaoka, F., 2001, DNA replication - SOS polymerases. *Nature* **409**: 33-34.

Hopfield, J.J., 1974, Kinetic proofreading: a new mechanism for reducing errors in biosynthetic processes requiring high specificity. *Proc. Natl. Acad. Sci. USA* **71**: 4135-4139.

Hunter, N., Chambers, S.R., Louis, E.J. and Borts, R.H., 1996, The mismatch repair system contributes to meiotic sterility in an inter-specific yeast hybrid. *EMBO J.* **15**: 1726-1733.

Lindahl, T. and Wood, R.D., 1999, Quality control by DNA repair. *Science* **286**: 1897-1905.

Matic, I., Rayssiguier, C. and Radman, M., 1995, Interspecies gene exchange in bacteria: The role of SOS and mismatch repair systems in evolution of species. *Cell* **80**: 507-515

Ninio, J., 1975, Kinetic amplification of enzyme discrimination. *Biochimie* **57**: 587-595.

Ninio, J., 1997, The evolutionary design of error-rates, and the fast fixation enigma. *Origins of Life and of the Biosphere* **27**: 609-621.

Orgel, L.E., 1963, *Proc. Natl. Acad. Sci. USA* **49**: 517.

Orgel, L.E,. 1970, The maintenance of the accuracy of protein synthesis and its relevance to ageing: a correction. *Proc. Natl. Acad. Sci. USA* **67**: 1476

Radman, M., 1999, Enzymes of evolutionary change. *Nature* **401**: 866-869.

Radman, M., Taddei,F. and Matic, I., 2000, Evolution-driving genes. *Res. Microbiol.* **151**: 1-5.

Radman, M., Taddei, F. and Matic, I., 2001, DNA repair systems and bacterial evolution. *Cold Spring Harbor Symp. Quant. Biol.* **65**: in press.

Radman, M., Wagner, R.E., Glickman, B.W. and Meselson, M., 1980, DNA methylation, mismatch correction and genetic stability. In *Progress in environmental mutagenesis*, (ed. M. Alacevic) Elsevier/North-Holland, Amsterdam, pp. 121-130.

Schubert, U., Anton, L.C., Gibbs, C., Nortbury, C.C., Yewdell, J.W. and Bennink, J.R., 2000, Rapid degradation of a large fraction of newly synthesized proteins by proteasomes. *Nature* **404**: 770-774.

Taddei, F., Radman, M., Maynard-Smith, J., Toupance, B., Gouyon, P.H. and Godelle, B., 1997, Role of mutator alleles in adaptive evolution. *Nature* **387**: 700-702.

Vogel, F. and Motulsky, A.G., 1996, *Human genetics: problems and approaches*, 3RD edition, Springer, Berlin.

Vulic, M., Dionisio, F., Taddei, F. and Radman, M., 1997, Molecular keys to bacterial speciation: DNA polymorphism and the control of genetic barriers. *Proc. Natl. Acad. Sci. USA* **94**: 9763-9767.

Wood, R.D., Mitchell, M., Sgouros, J. and Lindahl, T., 2001, Human DNA repair genes. *Nature* **291**: 1284-1289.

Towards a Memory-Based Interpretation of Proteome Data

JÁNOS MURVAI, KRISTIAN VLAHOVIČEK and SÁNDOR PONGOR
International Center for Genetic Engineering and Biotechnology, Padriciano 99, 34012 Trieste, Italy

1. INTRODUCTION

Understanding and managing genomic data has become a major bottleneck of biomedical research that calls for novel informatics approaches. The task is immensely complex: to "understand" the role of a protein for example implies inserting it into a host of interconnected and evolving frameworks of biological knowledge, including 3-D structures, molecular interactions, biochemical pathyways, genomic locations, spatial and temporal roles within the cell, the organism, the population and the species. *Similarity based predictions* play an important role in this process: similar biological functions or roles are mostly inferred from similar structure or similar molecular interactions, etc. This is usually carried out by comparing a protein sequence with a database of known sequences, using such programs as BLAST (Altschul *et al.*, 1990; Altschul *et al.*, 1997).

The ultimate goal of protein sequence comparison is to find biologically significant sequence similarities that can help one to infer the fold and/or the function of an unknown protein based on its sequence. The fundamental problem is that many of the biologically important sequence similarities are not significant in the statistical sense, i.e. the alignment scores or alignment patterns found between sequences similar in the biological sense can not be well distinguished from chance similarities. This is true both for protein domains that are shared by different protein superfamilies, as well as for complete sequences of proteins that carry the same function in different organisms (Figure 1).

Supramolecular Structure and Function 7, edited by Pifat-Mrzljak.
Kluwer Academic/Plenum Publishers, New York, 2001.

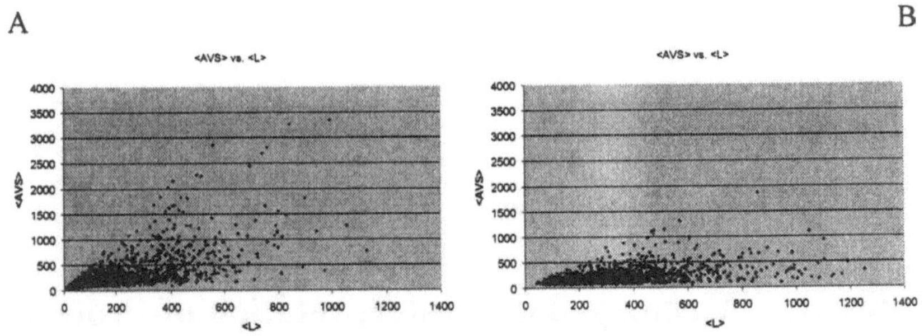

Figure 1. Self-smilarity of various sequence groups as determined by the BLAST program for A) protein domains (SBASE7.0, (Murvai *et al.*, 2000a)) and orthologous sequences taken from complete genomes (COG (Tatusov *et al.*, 2000)). Each sequence was compared with members of its own group using the BLAST program, and the average similarity score was calculated.

The intuitive notion of "biologicaly significant similarities" can be best described as membership in a similarity group characterized by established similarities, such as a group of domain sequences, or a group of orthologous proteins. For example, the similarity of a, say, trypsin-like domain to other trypsin-like domains is considered as "biologically significant", while its similarities to unrelated sequences are referred to as "chance". Prediction of domain homologies or of protein function is thus different from simple sequence similarity search in the sense that it is not only concerned with finding the most similar sequence to a query, but rather with assigning the query to a known similarity group. Also, statistical significance is allways understood in a context (a database, a scoring system and a theoretical model of chance similarities) but membership in a biologically significant similarity group (e.g. "trypsin-like") is generally used as permanent quality. A great deal of work has been invested into the modeling of sequence similarity groups in terms of Hidden Markov Models, sequence profiles and related techniques (Atwood, 2000). All these approaches provide *individual* descriptions to each group under study and consequently they are difficult to update with the stream of genomic data.

2. REPRESENTATION OF SIMILARITIES

The aim of the present work is to characterize the distribution of similarity scores within the known sequence similarity groups and to develop a general, exemplar-based prediction strategy on this basis. The examples shown in Figure 2 in fact suggest that the "self similarity scores" found between the members of various similarity groups (gray circles) are more or less separated from the chance or "non-self" similarities i.e. those that unrelated sequences have with members of the same group (black triangles).

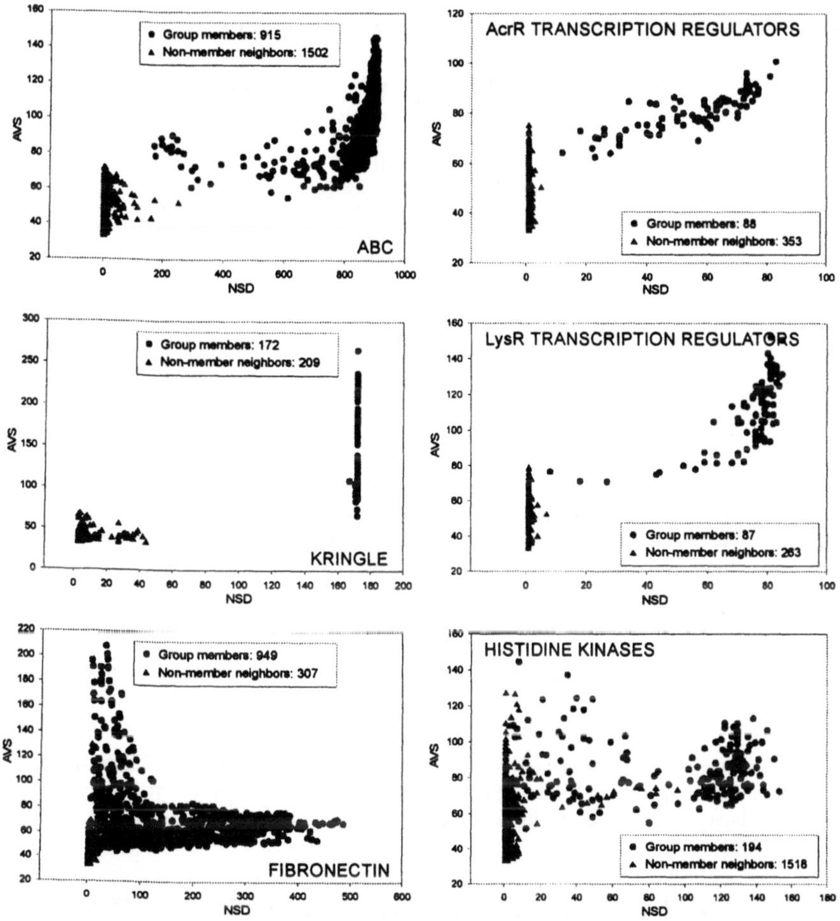

Figure 2. Number of significant similarities (NSD) vs. Average similarity score (AVS) plots in various domain sequence (SBASE) and orthologous protein sequence (COG) groups (Prosite, PFAM or COG reference given in parenthesis): ABC Transporter [PROSITE:PS00211]; Kringle [PROSITE:PDOC00020]; Fibronectin III domain [PFAM:PF00041]; Transcriptional regulators, AcrR family [COG:COG1309]; Transcriptional regulators, LysR family [COG:COG0583]; Sensory transduction histidine kinases [COG:COG0642]. The data are determined from a database vs. database comparison using BLAST. Each dot represents an SBASE domain or a COG sequence. The AVS and NSD values are computed from the BLAST similarities between the sequence and other members of the group ("self similarities", circles), or sequence and members of other groups ("non-self similarities", triangles)

Some groups are "tight" in the sense that most members are significantly similar to most other members, and, also there are differences in how well the self and non-self similarities are separated from each other. The intuitive concepts of "tightness" and "separation" can be visualized by plotting the empirical distributions the similarity scores as shown in Figure 3.

$$C = TP_{NSD} + P_{p}^{NSD}(NSD) + P_{nfp}^{NSD}(NSD) + TP_{AVS} + P_{p}^{AVS}(AVS) + P_{nfp}^{AVS}(AVS)$$

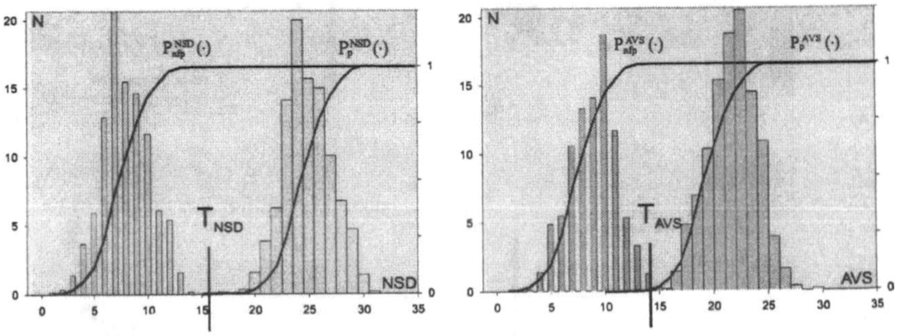

Figure 3. Variables used to characterize self- and non-self similarities. (Schematic histograms representation of the data shown in Figure 2). A) Empirical distribution of NSD of self-similarity scores $P_{p}^{NSD}(.)$ and non-self similarity scores $P_{nfp}^{NSD}(.)$ B) Empirical distribution of AVS of self-similarity scores $P_{p}^{AVS}(.)$ and non-self similarity scores $P_{nfp}^{AVS}(.)$ The four distribution curves are computed from a database vs. database comparison and stored for each sequence group. During prediction, a sequence is compared to the domain (or COG) database with BLAST, and NSD and AVS are computed for all categories present in the output. Then, the values of $P_{p}^{NSD}(NSD)$, $P_{p}^{AVS}(AVS)$, $P_{nfp}^{NSD}(NSD)$, and $P_{nfp}^{AVS}(AVS)$, are simply read out from the precomputed empirical distributions. The cumulative score C is the sum of these variables, plus two threshold premium values TP_{NSD} and TP_{AVS} that are have a value of 2.0 if *NSD* and AVS are above the group threshold T_{NSD} and T_{AVS}; respectively, and zero otherwise. (Murvai *et al.*, 2000b)

The approach presented here consists in characterizing the self- and non-self similarity distributions in a reference database, and then predicting group membership by comparing a query to the reference database. In this work we show results on domain sequences, using the SBASE domain sequence database (Murvai *et al.*, 2000b) and functional similarities, using the COG database of orthologous clusters (Tatusov *et al.*, 2000). The similarity scores are calculared with the BLAST program (Altschul et al., 1997). The key step of this approach is a simple transformation that represents a sequence object as a vector of similarity scores calculated with respect to a reference sequence database. This is equivalent of placing the sequence object into a similarity space with as many dimensions as there are members in the reference database. In this space, the object is no longer described in terms of its structure or sequence but rather by its similarities to the database objects. As the reference sequence databases can contain more than 100 thousand sequences, the similarity vectors would, in principle, have a very large number of components, even though most of them negligibly small. A solution to this problem is to filter the similarities by mathematical significance first, (with the BLAST program we achieve this by choosing a generously loose cutoff of P=0.8, ~ E=200), then to calculate group (class) averages of the similarity scores. The similarity of a sequence to a group *g* is then represented by two variables, the number of significant similarities, NSD_g, and the average of the corresponding similarity scores, AVS_g. This is equivalent of dividing the similarity space into small neighborhoods corresponding to each similarity group, and define a class-specific similarity function between the query and each of the sequence groups. Each neighborhood will center around one similarity group, and will contain members of the group as well as well as non-members that nevertheless display significant similarity to various members of the groups, as shown for several domain and COG groups in Figure 2 (a complete set of 2-D plots is available at www.icgeb.trieste.it). The variables used to describe each neighborhood are those shown in Figure 3. These are local variables in each neighborhood and characterize self-and the non-self similarity within the neighborhood (NSD_g, AVS_g). They are determined from a database vs. database comparison and stored for each neighborhood as a knowledge base of similarities.

3. CATEGORIZATION ALGORITHMS

The categorization is then carried out by first comparing a query sequence with the reference database, then evaluating the search output in comparison with this knowledge base using one of the 3 simple classification methods described below. The interesting point is that after this transformation, many of the most difficult sequence categorization tasks (i.e. those cases where the within-group similarity is is very far from what is considered significant in the statistical sense) can be quite efficiently solved using standard, simple non-parametric classification methods, of which we give three examples.

i) The first such approach is a nearest neighbor algorithm applied to the self similarities: The categories are ranked according the NSD and AVS_g values, and if the winner of both list is same, that category will be assigned to the query, other cases will be reported as "doubt". It is required, however, that the AVS_g and NSD_g values should be beyond threshold values defined as the smallest respective value found in the given group of the reference database (Figure 3). This nearest neighbor method provides correct predictions (no false positives or false negatives) in 80 % of the domain groups and 71 % of the functional groups. Examples are shown in the Table.

ii) A more efficient yet simple scoring method incorporates both self and non-self similarities The cumulative probabilistic score C_g described in Figure 3 (Murvai *et al.*, 2000b) does not require that either of the NSD_g or AVS_g values be beyond the respective thresholds; it will be calculated for all categories present in the search output and the category with the highest C_g value will be assigned to the query. It is used as a simple diagnostic tool, for a sequence to be considered a member of the group, its C score should reach the lowest value recorded for the group - other cases are reported as doubt. Table shows that the probabilistic score allows correct prediction in many further sequence groups, leaving incorrect predictions in only 112 domain groups (10%) and 194 functional groups (9 %).

iii) For the categories that can not be efficiently handled by either of the two previous methods we may apply more sophisticated techniques. We chose a standard feed-forward artificial neural network (ANN) architecture with 6 hidden units (Figure 4), using the 6 parameters shown in Figure 3 as the input (Murvai *et al.*, 2000c). The ANN method produces correct predictions in 55 domain groups and very low number of errors in the others (Table). At this stage, however, some of the weaknesses of the approach become apparent: artificial neural networks require a large training set, so they can not be applied efficiently to small domain sequence groups. This problem is especially conspicuous with the COG functional groups some of which are typically quite small. As a workaround, we selected homologous sequences from other databases (Swiss-Prot and PIR), but this is in some cases quite difficult because of the annotation conflicts that exist between the various databases. We developed ANNs only to 5 of the "problematic" functional sequence groups, and the results are as satisfactory as those obtained with the domain groups. While the parameters of the previous two methods are automatically derived from on a database vs. database comparison statistics, the neural network method requires an individual training for each of the sequence groups studied. The architecture is general, however, i.e. it does not depend on the domain group so the the training is automated. In addition, the trainig process takes a very short time (on the average it took less than 2 minutes for a domain group, using a 300 MHz single processor Sun work-station).

WWW servers for domain and function prediction are available at www.icgeb.trieste.it/sbase.

Name	Sequence group N[1,2]	NS[3]	AVS[4]	E[5]	Prediction method: Nearest neighbor correct	Nearest neighbor fp/fn	Probabilistic scoring correct	Probabilistic scoring fp/fn	Artificial neural network correct	Artificial neural network fp/fn	Final performance correct	Final performance fp/fn
SBASE release 7.0												
KRINGLE DOMAIN	61	172.92 (172-173)	156 (69.69-273.01)	1.27E-14	61	0/0	61	0/0	-	-	61	0/0
COLLAGEN TRIPLE HELIX REPEAT	211	879(878-879)	144.67(89.52-181.48)	3.82E-13	211	4/0	211	0/0	-	-	211	0/0
7 TRANSMEMBRANE RECEPTOR	825	650.84(23-798)	115.75(56.84-274.17)	2.10E-08	825	6/0	825	0/0	-	-	825	0/0
CYTOCHROME C OXIDASE SUBUNIT II.	233	294.66(17-406)	290.62(67-409.91	1.45E-33	233	1/0	233	0/0	-	-	233	0/0
ABC TRANSPORTERS	769	54.83 (6-104)	196.07 (74.62-440.93)	6.39E-20	769	11/0	769	6/0	769	1/0	769	1/0
SUSHI DOMAIN (SCR REPEAT)	165	287.10 (19-446)	73.12 (4407-155.45)	4.92E-03	165	30/0	165	3/0	165	0/0	165	0/0
PROTEIN-KINASE DOMAIN	1642	1456.05 (13-1621)	112.41 (54.36-273.00)	6.28E-08	1640	9/2	1642	30/0	1642	0/0	1642	0/0
TRYPSIN-LIKE DOMAIN	442	380.94 (9-419)	125.75 (47.94-353.36)	6.86E-10	442	170/0	442	3/0	442	0/0	442	0/0
ANK REPEAT	157	193.32 (5-447)	61.93 (46.43-90.24)	1.09E-01	156	33/1	157	4/0	155	2/2	155	2/2
FIBRONECTIN TYPE III DOMAIN	350	146.86 (5-489)	93.40 (54.05-267.33)	9.99E-06	350	514/0	350	8/0	350	2/0	350	2/0
EGF-LIKE DOMAIN	436	741.08 (3-1271)	53.86 (35-73.53)	1.60E+00	433	181/3	436	34/0	436	3/0	436	3/0
IMMUNOGLOBULIN DOMAIN	2584	1040.36 (4-1791)	120.66 (41.90-345.63)	1.20E-09	2584	3011/0	2584	633/0	2552	71/32	2552	71/32
Of a total of 52933 sequences					45117	7592/250	5775	2041/0	1165	166/99	52389	445/99
Of a total of 1515 of groups					1213	270/112	154	147/0	113	55/54	1427	90/54
COG												
Transcriptional regulators, LysR family	87	73.63 (8-85)	109.47 (70.92-152.38)	2.41E-07	87	0/0	87	0/0	-	-	87	0/0
Acyl-CoA dehydrogenases	58	48.20 (18-57)	139.07 (64.39-223.82)	2.72E-11	58	0/0	58	0/0	-	-	58	0/0
Transcriptional regulators, AcrR family	88	53.59 (12-83)	80.54 (62.56-100.81)	1.79E-03	86	0/2	88	0/0	-	-	88	0/0
Dipeptide/oligopeptide/nickel ABC-type transport systems, periplasmic components	62	48.25 (8-60)	112.80 (64.66-217.37)	1.55E-07	59	0/3	62	0/0	-	-	62	0/0
Glycosyltransferases I	107	48.20 (2-90)	86.27 (51.0-388.89)	5.51E-04	89	0/18	107	1/0	300	2/0	300	2/0
Sensory transduction histidine kinases	195	89.93 (1-153)	83.75 (46.0-144.62)	1.14E-03	145	1/49	194	6/1	264	2/4	264	2/4
Thiol-disulfide isomerase and thioredoxins	104	34.63 (1-72)	89.17 (48.0-156.0)	9.81E-05	79	0/25	104	4/0	218	1/6	218	1/6
Serine/threonine protein kinases	154	116.48 (4-146)	98.96 (59.88-147.58)	1.35E-05	130	0/22	152	2/2	1622	1/12	1622	1/12
Permeases	363	95.12 (1-238)	105.07 (38.5-447.17)	1.45E-06	288	0/75	363	0/5	428	29/16	428	29/16
Of a total of 28141 sequences					24728	71/3429	27457	684/684	888	35/35	27457	684/684
Of a total of 2104 of groups					1494	48/658	1910	194/194	5	5/4	1910	194/193

[1] The comparisons were done with the BLAST program

[2] Number of sequences in the group

[3] Number and range (max-min) of significant similarities to members of the same group

[4] Average similarity score and range (max-min) of significant similarities to members of the same group

[5] Expect value, calculated as $E = mn \cdot 2^{-S}$; $S = \dfrac{\lambda \cdot AVS - \ln K}{\ln 2}$, where m= 31411157, n= 100, k=0.138, lambda=0.318

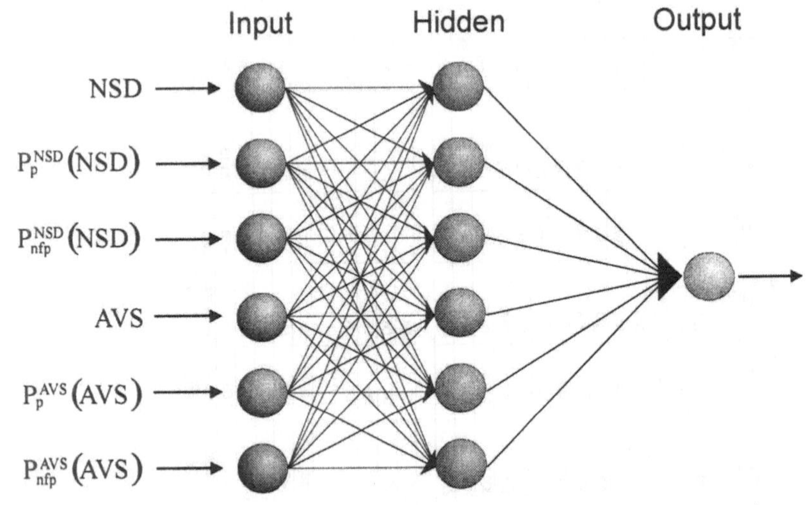

Figure 4. The backpropagation neural network architecture used for domain recognition. The variables are explained in the text and in Figure 3.

4. DISCUSSION

The predictive performance of these methods compares quite favorably with other methods such as Hidden Markov models (Sonnhammer *et al.*, 1997) and profile searches (Gribskov *et al.*, 1987). Essentially, the present approach provides efficient categorization in the time frame of a simple sequence similarity search, and does not even require iterative re/searching of the database. We believe that the success of this approach is primarily due to the similarity-based encoding of the data. This is apparent from the fact that artificial neural networks with only 6 input data could efficiently handle the categorization in all cases in which there was a sufficient number of data available for training.

We term the present method a memory-based approach, because its principles, especially the nearest neighbor algorithms *i* and *ii* are analogous in many respect to the memory-based learning paradigm described by Stanfill and Waltz (Stanfill and Waltz, 1986).Clearly, NSD, AVS and the probabilistic score C can be regarded as class-specific similarity (distance) functions that have parameters (thresholds, frequency distributions) which are learned from the database. Second, the reference database and the similarity knowledge base can in fact be considered as the memory of the system. This memory consists of two parts, a) Information expressed in terms of propositions added by human annotators (class labels, i.e. definition of domains or functions), and b) Information stored in inter-object connections that are determined automatically for each group from database vs. database comparisons (class average similarities, thresholds and and empirical distributions). It is noted that the simple, memory-based categorization strategies i) and ii) proved successful in the vast majority of the cases, neural networks had to be used for some of the very difficult and large groups. It is worth mentioning that similarity groups with low average statistical significance and short in sequence required more sophisticated methods (ii) and iii), the others could be handled by the simple, common sense nearest neighbor algorithm. It is important to point out the cases that could not be properly categorized by this approach: Out of 1515 domain groups, 17 such groups were found; they were not sufficiently separated for algorithms i and ii, but had too few data for training ANNs. On the other hand, out of 2104 COG clusters only 194 could not be categorized without mistakes using the simple algorithms; in those cases, however, annotation conflicts (between COG, Swiss-Prot and PIR) may have been the underlying reason. This points to the known fact that nearest neighbor type methods are quite sensitive to database errors (Ripley and Hjort, 1995).

The advantage of the current method is its speed and simplicity. Sequence classification tasks are typically solved by strategies that are based on individual models of the sequence groups described in terms of HMMs and profiles etc. These are computationally intensive strategies (as they require multiple alignments) and require a substantial human overhead for updating. The current method offers a comparable if not better performance (Table) moreover it is quite simple to update. Adding a new sequence group to the reference database consists in comparing the members of the new group to each other and to the database using BLAST, defining the 6 class specific quantities shown in Figure 3 (The two thresholds and the 4 P distribution curves) and updating the corresponding data of those classes that showed significant similarities with the new group. The speed of the method is a consequence of the encoding and data reduction schemes developed on the basis of earlier observations (Hegyi and Pongor, 1993; Murvai *et al.*, 1999). Grouping and evaluating the scores by classes means that only the immediate neighborhood of the similarity groups, i.e. a small fraction of the entire similarity space is explicitly represented, which however contains practically all detectable similarities (i.e. those with significance values $P<0.8$). Typically, a domain sequence will have significant similarities to 4 goups on the average, so the number of cases to be tested is not too large.

The present approach is conceptually different from most other sequence classification methods, because it uses an exemplar-based description (Smith and Medin, 1981) of the similarity groups. The classification methods used by the COG (Tatusov *et al.*, 2000) and the SYSTERS (Krause *et al.*, 2000) databases are also exemplar based, while most other methods use consensus-models that are probabilistic descriptions of the similarity groups (Smith and Medin, 1981). Exemplar-based methods have the important advantage that they are not statistically biased, i.e. allow the detection of atypical examples in the presence of an excess of known typical examples.

The present approach here is not limited either to the application examples or to the simple algorithms described here. All three algorithms were chosen because of their simplicity and none of them is considered as optimal in any sense; preliminary results show that adaptive kernel methods and support vector machine recognizers may provide useful alternatives (Ripley and Hjort, 1995). The choice of application areas will basically depend on the availability of a metrics distance (similarity measure) capable of distinguishing the object classes. It is important to note that the object classes used in this work were conceptual constructs based on human knowledge. It is equally possible to use classes automatically generated by clustering algorithms. Future application areas may include e.g. prediction of metabolic pathways, evolutionary networks and related classification schemes.

REFERENCES

Altschul, S. F., Gish, W., Miller, W., Myers, E. W. and Lipman, D. J., 1990, Basic Local Alignment Search Tool. *J. Mol. Biol.* **215**: 403-10.

Altschul, S. F., Madden, T. L., Schaffer, A. A., Zhang, J., Zhang, Z., Miller, W. and Lipman, D. J., 1997, Gapped Blast and Psi-Blast: A New Generation of Protein Database Search Programs. *Nucleic Acids Res.* **25**: 3389-402.

Atwood, T. K., 2000, The Role of Pattern Databases in Sequence Analysis. *Briefings in Bioinformatics* **1**: 45-59.

Gribskov, M., McLachlan, A. D. and Eisenberg, D., 1987, Profile Analysis: Detection of Distantly Related Proteins. *Proc. Natl. Acad. Sci.* **84**: 4355-4358.

Hegyi, H. and Pongor, S., 1993, Predicting Potential Domain Homologies from Fasta Search Results. *Comput. Appl. Biosci.* **9**: 371-2.

Krause, A., Stoye, J. and Vingron, M., 2000, The Systers Protein Sequence Cluster Set. *Nucleic Acids Res.* **28**: 270-2.

Murvai, J., Vlahovicek, K., Barta, E., Cataletto, B. and Pongor, S., 2000a, The Sbase Protein Domain Library, Release 7.0: A Collection of Annotated Protein Sequence Segments. *Nucleic Acids Res.* **28**: 260-2.

Murvai, J., Vlahovicek, K., Barta, E., Parthasarathy, S., Hegyi, H., Pfeiffer, F. and Pongor, S., 1999, The Domain-Server: Direct Prediction of Protein Domain-Homologies from Blast Search. *Bioinformatics* **15**: 343-4.

Murvai, J., Vlahovicek, K. and Pongor, S., 2000b, A Simple Probabilistic Scoring Method for Protein Domain Identification. *Bioinformatics*, in press.

Murvai, J., Vlahovicek, K., Szepesvári, C. and Pongor, S., 2000c, Prediction of Protein Functional Domains from Sequences Using Artificial Neural Networks. *Genom Research*, in press.

Ripley, B. D. and Hjort, N. L., 1995, *Pattern Recognition and Neural Networks*. Cambridge University Press. Cambridge.

Smith, E. E. and Medin, D. L., 1981, *Cathegories and Concepts*. Harvard University Press, Cambridge.

Sonnhammer, E. L. L., Eddy, S. R. and Durbin, R., 1997, Pfam: A Comprehensive Database of Protein Domain Families Based on Seed Alignments. *Proteins* **28**: 405-420.

Stanfill, C. and Waltz, D., 1986, Toward Memory-Based Reasoning. *Communications of the ACM* **29**: 1213-1228.

Tatusov, R. L., Galperin, M. Y., Natale, D. A. and Koonin, E. V., 2000, The Cog Database: A Tool for Genome-Scale Analysis of Protein Functions and Evolution. *Nucleic Acids Res.* **28**: 33-6.

Modelling DNA Structure from Sequence

KRISTIAN VLAHOVIČEK AND SÁNDOR PONGOR
International Center for Genetic Engineering and Biotechnology, Padriciano 99, 34012 Trieste, Italy

1. INTRODUCTION

Biological thinking in recent years has been profoundly influenced by the idea of local structural polymorphism in DNA. DNA is no longer considered as a featureless polymer but rather as a series of individual domains differing in flexibility and curvature. Local structural polymorphism is expected to contribute to the specificity of various biological events as gene regulation, packaging, for example, through regulating the affinity of protein binding (Travers and Klug, 1990). In contrast to traditional structural polymorphism (e,g, B, A or Z structures), here we deal with a localised micropolymorphism in which the original B-DNA structure is only distorted but is not extensively modified. The DNA segments involved in protein-induced and inherent DNA-bending are 10-50 base pairs in length (Olson and Zhurkin, 1996) i.e. they are longer than what can be easily handled by atomic resolution molecular modelling or quantum mechanical approaches. Traditional elastic models of DNA, that represent DNA as an ideally elastic, homogeneous cylindrical rod, were used to model macroscopic behaviour of long DNA segments, such as supercoiling (Langowski *et al.*, 1996; Olson, 1996). However, local DNA conformations and recognition by DNA-binding proteins are clearly sequence-dependent, so conventional elastic rod models of DNA, which do not explicitly represent the dependence of the elasticity on the base sequence, cannot say much about them. Here we attempt to briefly review advantages and limitations of the rod-models of DNA with particular regard to elastic modelling of local bending phenomena.

2. STATIC GEOMETRY MODELS

Rod models are the simplest form of DNA models; they picture DNA as a cylindrical rod of constant diameter. The shape in this case is the path or trajectory of the longitudinal axis, which can be either straight or curved. (Figure 1A). The common philosophy of rod models is to divide the rod into short cylindrical segments (e.g. of the size of a base pair, Figure 1B) and then to compute some rod parameter based on the segment parameters that

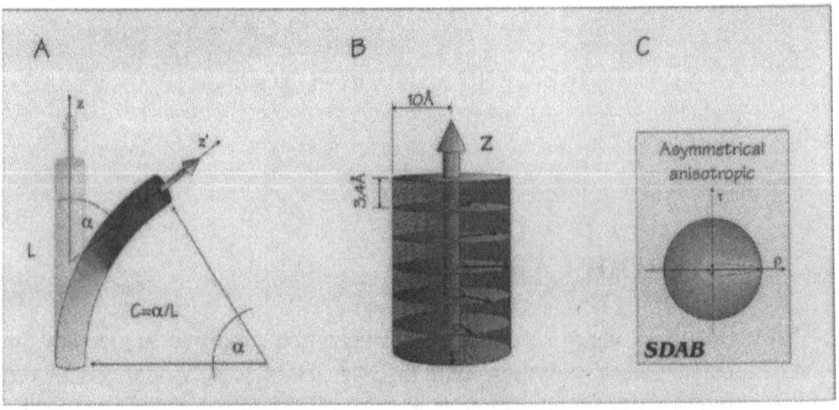

Figure 1. A: Curvature of an elastic rod. *B:* A segmented rod model. Each element corresponds to one base pair and may have different static deflection angles (roll, tilt twist) as well as different bending, torsional and stretching rigidity values. *C:* The sequence-dependent anisotropic bendability model of DNA. The arrows are proportional to the flexibility in a given direction (+x, -x, +y, -y). In the SDAB model, one direction (that of the major groove) is more flexible, the other three are more rigid.

have to be *a priori* known. Dinucleotide models define the base pair-size unit as one between two adjacent base pairs, which results in 16 different base pairs, or 10, if we allow strand symmetry (i.e. AA=TT). Trinucleotide models define the unit around the base pair in the middle of a given trinucleotide. This amounts to 64 or 32 different units, again depending on strand symmetry.

Static models are rigid rod models that only consider the static geometry of a constituting segment. Curvature in B-DNA was originally believed to be an attribute of A_n (n=4-6) tracts repeated in phase with DNA's helical repeat. Two static models have been proposed initially to explain the phenomenon. In the nearest neighbour model, the axial deflections of successive AA/TT dinucleotides add up to produce a curve (Trifonov and Sussman, 1980). In the so-called junction model, curvature is produced as the modified B-DNA structure of A_n tracts joins with adjacent B-DNA (Wu and Crothers, 1984).

More recently it became clear that in addition to the A_n tracts, DNA curvature involves additional sequence elements (Brukner *et al.*, 1993; Satchwell *et al.*, 1986), and more sophisticated models were proposed which included wedge angles for all 16 dinucleotides (Bolshoy *et al.*, 1991; De Santis *et al.*, 1990).

Current static models consider dinucleotide geometries derived from direct measurements such as X-ray crystallography (Bansal *et al.*, 1995; Olson *et al.*, 1995) and NMR (Gabrielian and Pongor, 1996; Ulyanov and James, 1995), from statistics of nucleosome binding data (Goodsell and Dickerson, 1994; Satchwell *et al.*, 1986) or from gel electrophoretic analysis of concatenated synthetic oligonucleotide repeats (Bolshoy *et al.*, 1991; De Santis *et al.*, 1990). The parameters explicitly considered are Roll, Tilt and Twist angles, and all other parameters are considered equal to that of B-DNA (the diameter of the rod plays no role in this calculation). Given the geometry of the segments one can calculate the trajectory of DNA by adding the distortions of the successive segments. This exercise is only seemingly simple, because addition of rotation operations like roll, tilt and twist are not commutative, so the results in principle depend on the serial order of the operations (el Hassan and Calladine, 1995). The resulting error is negligible for short DNA segments and for small angles (and generally speaking, roll and tilt angles are small). The calculation will however give large uncertainties for long segments or for large distortions. There are a number of programs available that can calculate an approximate DNA trajectory starting from a DNA sequence (Bansal *et al.*, 1995; De Santis *et al.*, 1990; Dlakic and Harrington, 1998; Shpigelman *et al.*, 1993; Wheeler, 1993). Table 1 shows that the current dinucleotide models give good qualitative correlations with the known experimental values, i.e. curved and straight motifs can be distinguished even though all models mispredict some of the motifs. It has to be mentioned that gel mobility anomaly of DNA that is used to validate the models is itself strongly dependent on environmental factors such as metal salts and temperature, so our knowledge on straight and curved motifs is qualitative rather than quantitative.

3. SIMPLE ELASTIC MODELS

Elastic bodies are those that regain their original shape after the stress is removed and in which the displacement (distortion) is proportional to the force applied. Single-molecule experiments show that these conditions are in fact met for small distortions of DNA molecules. If a DNA rod is ideally elastic one can compute the energy necessary for bending, stretching or torsional deformation (Barkley, 1979). For example, the energy required to bend the rod of length *L* to a given curvature can be calculated as

$$\Delta G = \frac{1}{2} EIL\alpha^2 \tag{1}$$

Here α is the curvature as defined in Figure 1A and *I* is the moment of inertia that, for a cylindrical rod of r radius is given as $\pi r^4/4$. *E* is the Young's modulus. Based on a variety of physico-chemical measurements and direct molecular stretching studies, the Young's modulus of average DNA is about 3.4 x 10^8 N/m^2, a value close to that of polypropylene or phenol resins.

By "simple elastic models" we mean those that consider DNA as an originally straight cylindrical rod with one *homogeneous isotropic elasticity* parameter (for extensive reviews see Langowski *et al.*, 1996; Olson and Zhurkin, 1996; Vologodskii and Frank-Kamenetskii, 1992). This means on the one hand that the model is not sequence dependent, i.e. all segments are equal, and on the other, the model is equally bendable (deformable) in all directions. The phenomena that can be described with such a simplified elastic model include the gross shape changes of DNA like supercoiling, response of plasmids to stress, etc. The nature of the answer is qualitative, for example one can state that the shape of a plasmid-like elastic ring model adopts, in response to torsional stress, shapes that are reminiscent of the shapes observed for supercoiled plasmids. This fact shows that some properties of DNA are in fact reminiscent of those of rubber strings, i.e. depend mostly on properties that are common to simple mechanical systems. The underlying, very complex mathematics can be performed by *finite element analysis*, a technique routinely used for the analysis of stresses and displacements in engineering (Zienkiewicz, 1991). This technique has been applied with success for small DNA displacements (Bauer *et al.*, 1993; Yang, 1993). Even though simple elastic models do not contain local (i.e. sequence dependent) information, they allow on the other hand to demonstrate how local structures might act on the elastic behaviour of the entire molecule. For example, incorporating a fixed curve into a DNA model will influence the gross shape of plasmids (Zhang *et al.*, 1994) and will bring distant sites into each other's vicinity (Rippe *et al.*, 1995). One can subjectively conclude that the isotropic elastic models can, efficiently model an impressive range of phenomena connected with the propagation of stress along the molecule.

4. SDAB: A SEQUENCE DEPENDENT ANISOTROPIC MODEL OF DNA BENDABILITY

For the modelling of local phenomena sequence dependence has to be incorporated into the elastic models. It is also known that bending is anisotropic (Schellman, 1974; Zhurkin *et al.*, 1979), i.e. DNA bends much more easily towards the major groove than in any other direction, moreover, bending in the tilt direction is not favourable in energetic terms (for a review see Olson and Zhurkin, 1996). We have developed anisotropic bendability parameters using the enzyme DNaseI (Brukner *et al.*, 1995). This enzyme bends DNA towards the major groove and binds virtually to all DNA sites without pronounced sequence specificity. DNaseI cutting rates can thus be used as an estimate of DNA bendability, which can in turn can be scaled to an approximate DNA rigidity scale (Gromiha *et al.*, 1996). The result is a simplified segmented rod model, shown in Figure 1B. In this *sequence dependent anisotropic bendability* (*SDAB*), each disk corresponds to one base pair and the arrow indicates the direction of facilitated flexibility (i.e. that of the major groove). In principle, such an anisotropic bendability model can have different bending flexibility in all directions. As a simplification, we take bendability towards the major groove as the principal parameter, and consider DNA 10-times stiffer in all other directions (see sketch in Figure 1C). The model does not include static deflection components, i.e. similar to isotropic models, it considers DNA as an originally straight rod. Neither does the model incorporate torsional flexibility. In other terms it is a stripped down model that is designed to reflect one aspect: local bending phenomena. In contrast to the isotropic elastic models, the SDAB model is non-linear in terms of displacement response. The non-linearity of the SDAB model is due to two factors: i) Material non-linearity is the non-linear bending characteristics of each base pair, that we term bending anisotropy; ii) Geometric non-linearity occurs in the case of non-negligible displacements. Finite element analysis is very appropriate to study such non-linear problems.

The SDAB model pictures DNA as an anisotropically bendable rod whose bendability is deduced from DNaseI experiments. However, the concept bendability was used also previously, by Travers and co-workers who determined the frequency of trinucleotides on the inside and outside faces of DNA bent around the nucleosome particle. These frequency data were then considered as a measure of bendability (without an explicit model similar to the one outlined here). The ranking of trinucleotides is however slightly different according to the DNaseI and nucleosome scales, respectively.

The first attempt to incorporate the nucleosome frequency data into a physical model was that of Goodsell and Dickerson who devised a static rod model in which the roll angles were derived from re-scaled nucleosome frequency data. This model was quite efficient in predicting, e.g. GC-type curvature, and Goodsell and Dickerson designed a simple and fast algorithm, BEND, for the prediction of curvature. This gave us the idea to use the DNaseI data in an analogous manner, using the BEND algorithm. In the following sections we apply the SDAB model first to 3 problems: i) estimation of bending energies and their correlation with observable quantities (sequence-dependent curvature, protein-DNA binding free energy). ii) Modelling of vibrational phenomena and finally iii) Prediction of curvature from sequence

5. SDAB: CALCULATION OF BENDING ENERGY

Modelling of minicircles of curved and straight DNA (Gabrielian *et al.*, 1998). A simple experiment can show the macroscopic anisotropy of the SDAB model (Figure 2). A rod model is bent to the same extent into various directions as shown in the inset, and the bending energy is determined by finite element methods and plotted against the direction of bending (angle β) Sequences that are repeats of curved DNA motifs, in fact, will have a rotational preference, i.e. there will be one stable energy minimum. This means that such a rod-model has a preferred direction of bending, so as a result of thermal fluctuations it will preferentially bend into one direction. In other terms, the physically measurable average conformation of such a model will be curved. The straight motifs, on the other hand, have either no minima or have two minima in opposite angular directions.

Another consequence of the energy minimum found in the circles is that, in the minimum energy conformation of helically phased repeats, certain motifs will face inward and the others outward. For example, in repeats of AAAAGGGCCC, the major groove of the GGGCCC motif faces inwards and the major groove of AAAAA are on the outer side of the circle, as found, for example in nucleosomes. The roll values at the central GC are the highest while there are slight negative rolls in the AAAAA tract Figure 2b. As the roll/twist profile of the lowest energy conformation illustrates, the shape of the circle is reminiscent of a polygon - in fact quite similar to those recently obtained for the same molecule using AFM microscopy (Han *et al.*, 1997). This means that the lowest energy conformation has slight kinks, which is reminiscent of the mini-kink model postulated by Zhurkin and co-workers (Zhurkin *et al.*, 1979). The high-energy conformers on the contrary, have smaller kinks (Figure 2C).

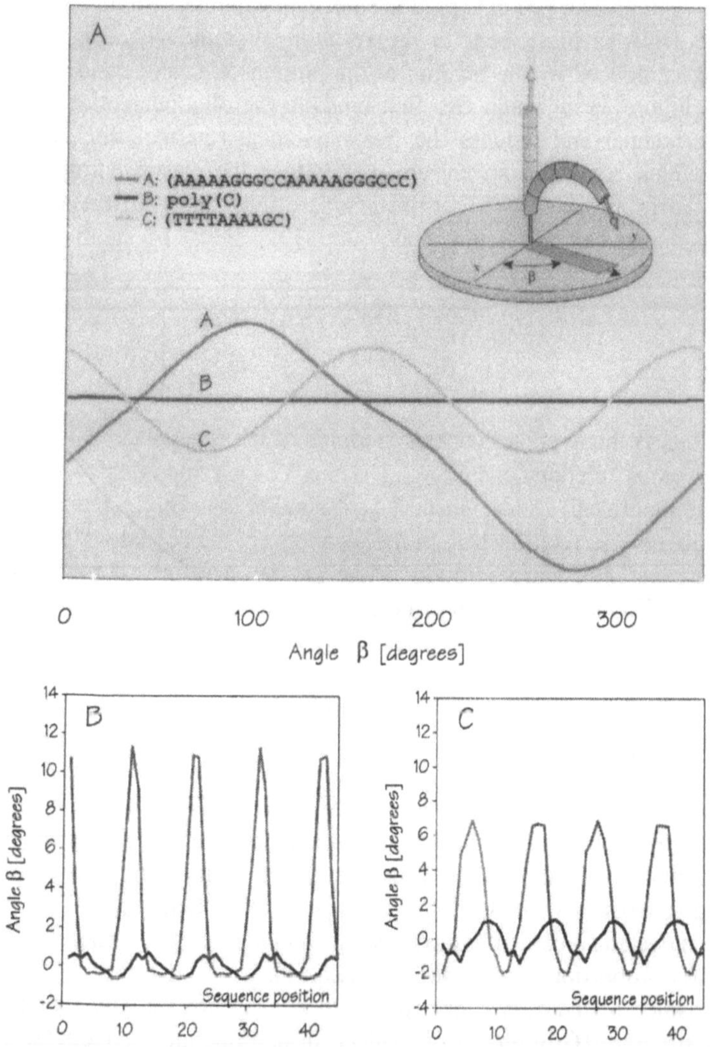

Figure 2. Testing bending anisotropy in DNA by finite element methods30. A DNA model was built (as outlined in Fig. 1c) from the repeat motifs shown. The models were bent to a given curvature in various directions [angle β in *A*] and the bending energy of the model was calculated for each direction by finite-element methods. Curved motifs (e.g. motif 1) exhibit a single energy minimum, which corresponds to a bending preference in one direction. Straight motifs have either no minimum (motif 2) or minima in two opposite directions (motif 3). The calculations were carried out using the Cosmos/M program (release 1.75A); the curvature value was uniformly 3.43° per base pair, and the energy was normalised to one base-pair unit, for better comparison (Gabrielian *et al.*, 1998).

Protein/DNA binding: DNA rigidity versus complex stability (Gromiha *et al.*, 1997) Repressor proteins bind to short DNA motifs with high specificity and the DNA is often bent in the resulting protein/DNA complex since bending is induced by the binding of the protein. We can consider a simple model (Figure 3) in which Cro first binds to the oligonucleotide in a non-specific manner and reduces the free movement (thermal fluctuations) of DNA, which results in an entropy loss. Since the elastic entropy can be calculated from the $<\theta^2>^{1/2}$ root mean square fluctuations of the model, the entropy change can be calculated as

$$\Delta S = nR\ln\left[\frac{\langle\theta^2_{bound}\rangle^{1/2}}{\langle\theta^2_{free}\rangle^{1/2}}\right] = nR\ln\left[\frac{E_{free}}{E_{bound}}\right] \qquad (2)$$

where E_{free} is the average Young's modulus of the segment, n is the number of degrees of freedom and E_{bound} is the Young's modulus of the bound (quasi immobilised) DNA. Since E_{free} is smaller than E_{bound}, this equation shows an adverse relationship. In the second step, the cognate will bend to a curvature of 78.5 degrees found in the crystal structure of the Cro/DNA complex. This second step will take place only for cognate DNA. Takeda and co-workers have measured a set of free energy data for cognate and non-cognate DNA that showed a puzzling diversity of values. If we now calculate the free energy values according to the simple model outlined here (Figure 3) the cognate and non-cognate molecules will clearly separate into two groups, that in turn clearly follow the tendencies predicted by equation 1 and 2. Namely, cognate (operator) and non-cognate (non-operator) DNA follow two adverse, quasi-linear relationships. In the operator sequences, DG is higher for stiffer molecules, i.e. the stiffer the molecule, the weaker the binding (energy is wasted on bending a stiff oligonucleotide). This in fact can be expected since Cro has to curve the molecule, and the energy required is linearly proportional to the stiffness (Eqn 1).

The slope of the curve corresponds to a curvature actually found in the crystal structure (Gromiha *et al.*, 1997). In non-operator sequences, on the other hand, ΔG is lower for stiffer sequences, i.e. the stiffer the sequence, the stronger the binding. Namely, relatively less energy is needed for restricting a rigid molecules movement.

6. SDAB: MODELLING VIBRATIONAL PHENOMENA IN DNA

The interest in vibrational properties of DNA is understandable, as it has

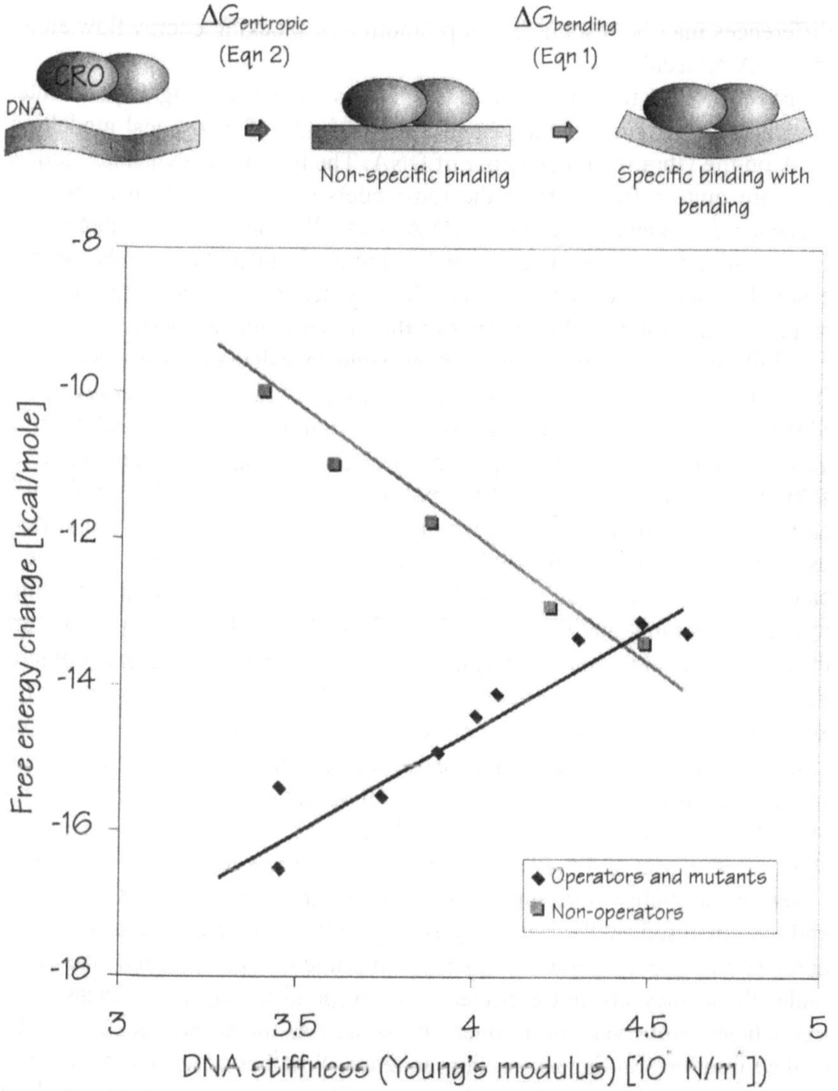

Figure 3. A simple model of Cro/DNA interactions (above); Binding of Cro operator to oligonucleotides of different stiffness (below). Relationship between the average stiffness (Young's modulus) of DNA and the free energy change (ΔG) of operator (diamonds) and non-operator (squares) DNA sequences (Gromiha *et al.*, 1997).

been speculated for a long time that energy can propagate along the chain not only in terms of stress (like in supercoiling or bending) but also as waves in a vibrating string. Consequently sequence dependent vibrational differences may be responsible in promoting or blocking energy flow along the DNA molecule.

Even though to our knowledge there is no convincing experimental evidence to this supposition, there is a vast body of theoretical modelling work on the vibrational behaviour of DNA. The non-linear dynamics models used are quite different from the rod-models and usually do not consider sequence dependence. Moreover, DNA is usually considered as vibrating *in vacuo* and effects like solvation and binding of molecular ligands can not easily be taken into consideration. In this section we show preliminary results on the vibrational properties of the SDAB model of DNA.

Finite element methods make it possible to calculate natural vibration frequencies and/or dynamic response of a mechanical model. However, the SDAB model is a non-linear case since the bending flexibility is different in various directions. There are possibilities to derive approximate analytical solutions for the calculation of normal modes in this particular ("bilinear") case, however numerical solutions are also straightforward and are widely used in the vibrational analysis of mechanical structures. In these calculations, the mechanical model is first exposed to an initial bending impact or bending, and then it is left to vibrate by itself. The frequencies are then determined by Fourier analysis of the displacements in the time domain. These vibrations are calculated for a model in vacuo. The surrounding environment can be modelled as a viscous medium that damps the vibrations - in principle damping is not expected to influence very much the frequencies as the damping ratio has small values. Figure 4 shows a simple modelling experiment. A model of 32 residues was fixed on one end as shown in the inset and the free end was deflected to a moderate curvature of 7 degrees per helical turn and then left to vibrate The movement of the free end was analysed by Fourier analysis to give the frequency spectrum shown here. The frequency values in such an experiment depend on the size of the molecule so they are not expected to correspond to physically measurable frequencies. However, the frequencies as well as the amplitudes (intensity) of the vibrations do depend on the sequence. Whether or not the differences are meaningful in the physical sense, will have to be determined by experimental work.

7. SDAB MODEL: PREDICTION OF BENDABLE AND CURVED SEGMENTS

The single energy minimum in Fig. 2 is a consequence of two facts: the base pairs are themselves anisotropically bendable and, moreover, they are in a sequence that dictates that the direction of principal bendability is along

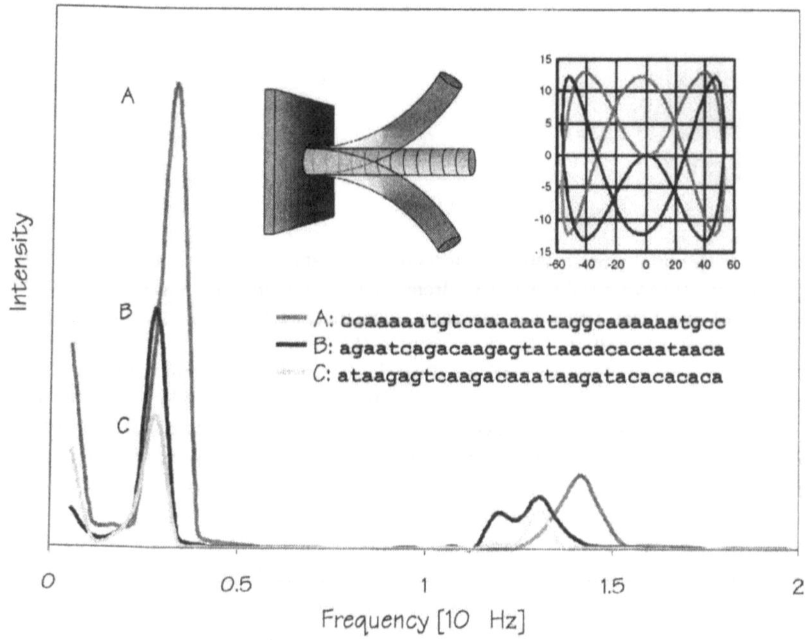

Figure 4. Sequence-dependent vibrations of DNA-models. A is an inherently curved sequence from *Leishmania tarentolae*, B and C are random shuffled variants of the same sequence, with no predicted curvature.

one face of the DNA helix. This can be best illustrated by a vector representation in helical circle diagrams - technique originally developed for amphipathic α-helices in proteins32 (Fig. 5A). In these diagrams, the bendability value for each base pair is plotted as a vector pointing towards the major groove. In fact, inherently curved motifs show asymmetrical distribution of bendability; large vectors are on one side (Fig. 5A). The vectors for base pairs in straight sequences, by contrast, have a rather symmetrical distribution. This amounts to saying that in randomly chosen DNA the thermal fluctuations in various directions will cancel out, while inherent bending is dependent upon a specific sequence. It is interesting to note that there is a mathematical analogy between bent DNA and amphipathic α-helices in proteins. Helical circle diagrams were originally developed for amphipathic α-helices in proteins. Furthermore, the vectorial average shown in the figures is analogous to the hydrophobic moment of Eisenberg (Eisenberg et al., 1984).

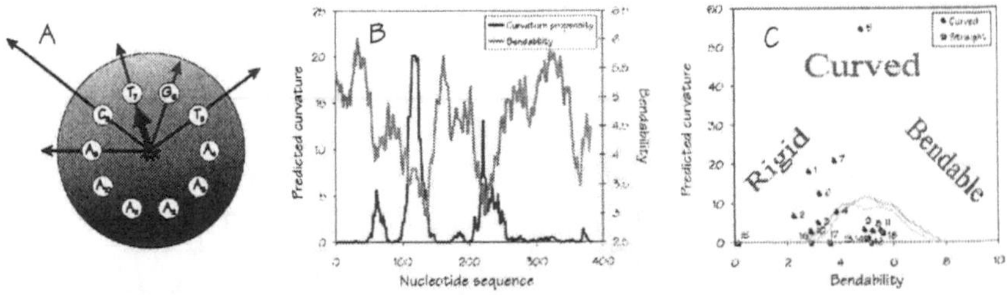

Figure 5. A: Vectorial representation (helical circle diagram) of DNA bendability in a curved sequence motif [(A)AAATGTCAAA(A)] from a *Leishmania tarentolae* class II minicircle. The length of each black arrow is proportional to the magnitude of the bendability parameter at a given sequence position. The gray arrow is the vectorial average of the bendability vectors and is considered to be a measure of predicted curvature. *B:* Predicted curvature and bendability against sequence (accession code Genbank LEIKPMNC2) plot. The values are calculated for 32-bp-sequence windows and plotted at the starting point of each window. *C:* Predicted curvature against bendability plots of curved and straight sequences from the Table 1. The lines indicate the border of random sequences obtained by random shuffling of the sequences of *Haemophilus influenzae* genome and yeast chromosome III.

Given the fact that the bendability of the *SDAB* model is asymmetrical towards the major groove, thermal fluctuations will result in bending of the model. How can one calculate the curvature from these values? Gabrielian *et al.* (Calladine and Drew, 1996) used the length of the vectorial average (as shown in Fig. 3) as a measure of predicted curvature; however, more rigorous geometry calculations can also be used: bendability values (see Table 1) can be considered to be proportional to (but not necessarily identical with) static trinucleotide roll values. In other words, the bendability parameters can be considered to be analogous to a static geometry model in which the dynamic contribution of thermal fluctuations is included.

With this assumptions, the *SDAB* model gives accurate predictions of DNA curvature that compare quite favourably to those of the static geometry models (see Table 1). Table 1 also includes figures calculated using the so-called consensus bendability scale14, which was developed in order to increase the sensitivity of the prediction towards GC-based curved motifs - these are often mispredicted using both the static dinucleotide models15 and the original DNaseI scale.

One application of this prediction is to analyse long DNA sequences, segment by segment, and then to visualise the distribution of predicted curvature (Fig.5B). We have developed simple visualisation methods in which several sequence-dependent properties can be combined, either as 1D-plots (along the sequence). In a 2D plot two properties of a segment (like average bendability and predicted curvature) are plotted against each other so each segment of a genome will produce a point in the plane (Fig. 5C). The distribution of these points is genome specific, and also it allows one to

Table 1. Analysis of curved and straight sequence motifs with various methods

No.	Sequence motif	Static geometry model				Sequence dependent anisotropic bendability (SDAB) model			
		Curvature[°/helical turn]				Bendability [a.u.]		Predicted curvature [°/helical turn]	
		Electro-phoresis	X-rays	NMR	Nucleo-some	DNaseI	Consen-sus	DNaseI	Consen-sus
		Curved DNA							
1	$(aaaattttgc)_n$	26.2	6.9	18.3	13.7	2.8	2.4	21.1	17.4
2	$(aaaattttcg)_n$	21.0	3.8	3.8	17.7	2.2	2.3	16.6	17.2
3	$(tctcaaaaaacgcgaaaaaaccggaaaaaagc)_n$	27.1	8.2	16.7	17.1	3.2	3.2	15.4	15.9
4	$(ccgaaaaagg)_n$	14.7	6.8	13.1	23.3	3.9	4.3	17.1	20.2
5	$(tctctaaaaaatatataaaaa)_n$	27.8	3.0	7.5	10.9	4.5	3.2	27.6	18.5
6	(ggcaaaaaac)n	26.8	12.0	20.1	20.4	3.2	3.3	19.2	19.5
7	ccaaaaatgtcaaaaaataggcaaaaaatgcc - *L.tarentolae* kinetoplast	26.0	6.4	15.7	19.6	3.9	3.5	21.8	20.5
8	aaaaactctctaaaaactctccctagaggggccctagagggc	19.4	7.8	10.6	13.3	5.2	4.6	13.6	12.8
9	aaaaactctctaaaaactctctagaggggccctagagggccc	17.5	7.2	13.1	16.4	5.1	4.6	13.3	13.2
10	agaattgggacaaaaattggaaatttttaaggg - *C.risortia* bent sat. DNA	18.5	8.9	8.7	12.4	3.3	3.0	13.5	12.6
11	$(aaaaactctctaaaaactctcgaggggccctagagggcccta)_n$	21.6	5.2	10.1	14.5	5.0	4.7	13.8	10.8
		Straight DNA							
12	$(atctaatctaacacaacaca)_n$	0.8	0.5	2.7	1.2	5.1	4.4	0.8	0.8
13	actacgttaaatctatcaccgcaagggataaa - OR3 operator region	10.4	5.5	4.9	5.9	5.0	4.4	10.4	7.8
14	actacgttaaatctatcaccacaagggataaa - OR3 region, mutated	11.0	5.5	3.4	6.2	4.9	4.4	10.6	8.1
15	$(a)_n$ - poly-A[2]	0.008	0.000	0.008	0.000	0.100	0.063	0.002	0.002
16	$(ttttaaaacg)_n$	1.5	7.1	14.5	10.7	2.8	2.5	2.3	4.2
17	$(ttttaaaagc)_n$	1.7	0.8	16.0	16.3	3.6	3.3	3.0	9.6
18	$(aaaaactctctaaaaactctcgggccctagaggggccctaga)_n$	27.1	3.1	5.9	7.5	4.9	4.6	12.8	8.3

[1]The angular deflection [°/10.5bp hel.turn] was calculated with the bend.it server. (The curvature units of Trifonov et al. correspond to 4.5°/bp or 47.25°/10.5 bp hel.turn)
[2]By definition, homopolymers should give zero curvature. The non-zero value indicates the numeric precision of the calculation.

visualise "outliers" i.e. segments and regions that are different from the rest of genome. For example, protein coding and non protein coding regions show quite different distributions, extremely curved, rigid or flexible segments are much more frequent in the letter than in the former (Fig. 6). Several of these calculation methods are now included into an Internet-based server http://www.icgeb.trieste.it/dna

A second, related, application of the method is to visualise curved DNA molecules. Once such a potential motif is identified in the sequence, the experimenter may wish to see what it looks like in three dimensions. One possibility - now included in the www server - is to reconstruct atomic co-ordinates from the idealised trajectory calculated by a simplified rod model. Standard modelling programs can be used to generate the atomic co-ordinates which then have to be refined with a combination of energy minimisation by restricted simulated annealing (molecular dynamics) and geometry optimisation. The latter steps are necessary because introduction of bends into a DNA molecule leads to atomic clashes (especially in the backbone). This method accurately reproduces the trajectory predicted by the rod model and the phosphate atoms are reproducibly re-positioned by the simulated annealing procedure. However, the accuracy of these predicted atomic models is of course limited and in our view they can be used mainly for the visualisation of the sidedness of the curve, i.e. localisation of the major and minor grooves as well as sequence motifs with respect to the curvature.

8. CONCLUSIONS AND FUTURE DIRECTIONS

We can conclude that static and dynamic rod models describe quite different aspects of the DNA molecule with as few parameters as possible. If they succeed - and they do, quite surprisingly - it means that the parameters used by the model sufficiently explain a given aspect of the molecule's behaviour. Both type of models can predict curvature in short DNA segments - in this respect they can be considered equivalent (Gabrielian *et al.*, 1996) The differences in predictive accuracy may be due to the parameterisation (e.g. trinucleotide vs. dinucleotide representation, electrophoresis vs. nucleosome data, etc.) and not to the models themselves. Considering the significance of these structural features, it is worth to mention that bendability/curvature characteristics are conserved in evolutionarily (Calladine and Drew, 1996; Gabrielian *et al.*, 1998) and functionally related sequences (Schatz and Langowski, 1997), and they

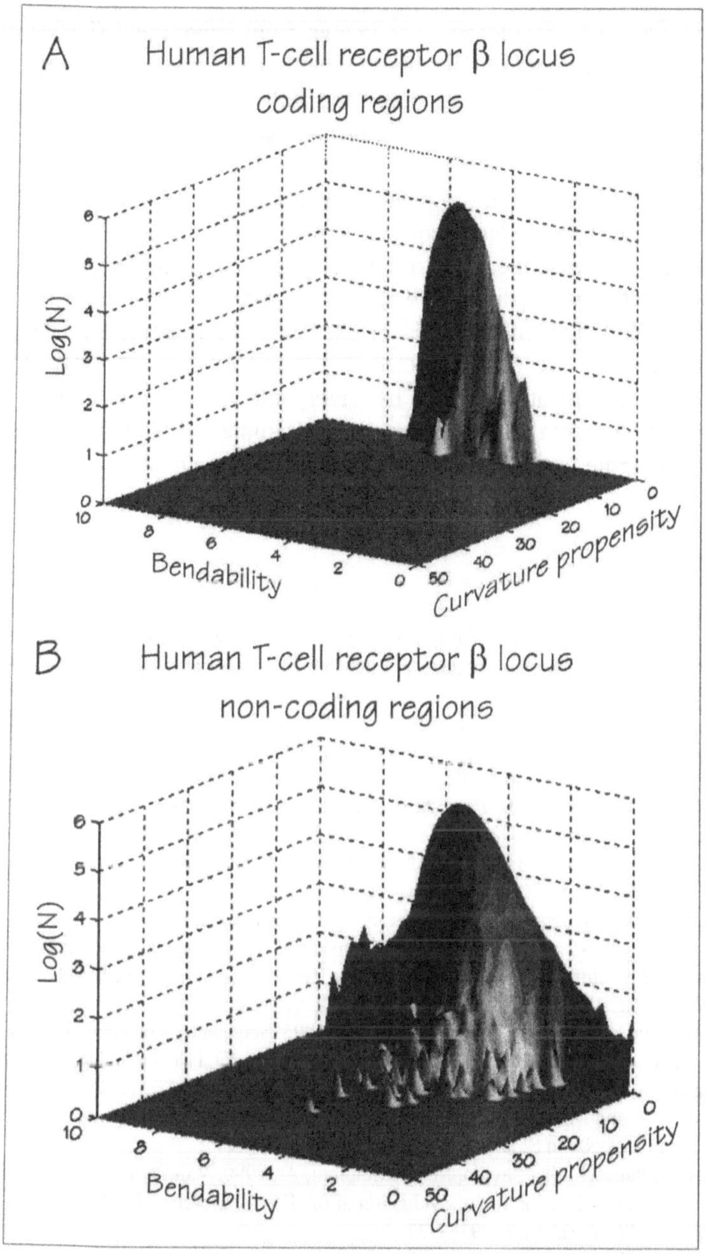

Figure 6 3D Curvature propensity vs. bendability histogram of the Human T-cell receptor locus (Genbank: humtcrb) The protein coding regions *A* have a tight distribution around average bendability and low curvature. The non-protein-coding regions *B* have a higher number of stiff (low bendability), flexible and curved segments.

correlate well with regulatory sites known from experiment (Langst *et al.*, 1997). So one can predict that these methods will have a continued role in locating functional sites in genomes and will hopefully contribute to our understanding how the putative conformational signals may operate at the genomic level.

Refinements of rod models, especially the incorporation of tetranucleotide-based description will probably increase the predictive accuracy and increase the scope of applications. Future models may combine both static and dynamic features with particular reference to the principles of non-linear DNA dynamics (Yakushevich, 1994). At the same time molecular mechanic models are also expected to expand to a wider range of local phenomena as illustrated recently by Lavery and associates (Sanghani *et al.*, 1996). In the foreseeable future however the simple rod models will continue to play a key role in the large-scale analysis of genomic data.

REFERENCES

Bansal, M., Bhattacharyya, D. and Ravi, B., 1995, Nuparm and Nucgen: Software for Analysis and Generation of Sequence Dependent Nucleic Acid Structures. *Comput Appl Biosci.* **11**: 281-7.

Barkley, M. D., Zimm, B., 1979, Theory of Twisting and Bending of Chain Macromolecules: Analysis of the Fluorescence Depolarization of DNA. *J. Chem.Phys.* **70**: 2991-3007.

Bauer, W. R., Lund, R. A. and White, J. H., 1993, Twist and Writhe of a DNA Loop Containing Intrinsic Bends. *Proc. Natl. Acad. Sci. USA* **90**: 833-7.

Bolshoy, A., McNamara, P., Harrington, R. E. and Trifonov, E. N., 1991, Curved DNA without a-A: Experimental Estimation of All 16 DNA Wedge Angles. *Proc. Natl. Acad. Sci. USA* **88**: 2312-6.

Brukner, I., Dlakic, M., Savic, A., Susic, S., Pongor, S. and Suck, D., 1993, Evidence for Opposite Groove-Directed Curvature of Gggccc and Aaaaa Sequence Elements [Published Erratum Appears in Nucleic Acids Res 1993 Mar 11;21(5):1332]. *Nucleic Acids Res.* **21**: 1025-9.

Brukner, I., Sanchez, R., Suck, D. and Pongor, S., 1995, Sequence-Dependent Bending Propensity of DNA as Revealed by Dnase I: Parameters for Trinucleotides. *Embo J.* **14**: 1812-8.

Calladine, C. R. and Drew, H. R., 1996, A Useful Role for "Static" Models in Elucidating the Behaviour of DNA in Solution. *J. Mol. Biol.* **257**: 479-85.

De Santis, P., Palleschi, A., Savino, M. and Scipioni, A., 1990, Validity of the Nearest-Neighbor Approximation in the Evaluation of the Electrophoretic Manifestations of DNA Curvature. *Biochemistry* **29**: 9269-73.

Dlakic, M. and Harrington, R. E., 1998, Diamod: Display and Modeling of DNA Bending. *Bioinformatics* **14**: 326-31.

Eisenberg, D., Schwarz, E., Komaromy, M. and Wall, R., 1984, Analysis of Membrane and Surface Protein Sequences with the Hydrophobic Moment Plot. *J. Mol. Biol.* **179**: 125-42.

el Hassan, M. A. and Calladine, C. R., 1995, The Assessment of the Geometry of Dinucleotide Steps in Double-Helical DNA; a New Local Calculation Scheme. *J. Mol. Biol.* **251**: 648-64.

Gabrielian, A. and Pongor, S., 1996, Correlation of Intrinsic DNA Curvature with DNA Property Periodicity. *Febs Lett.* **393**: 65-8.
Gabrielian, A., Simoncsits, A. and Pongor, S., 1996, Distribution of Bending Propensity in DNA Sequences. *Febs Lett.* **393**: 124-30.
Gabrielian, A., Vlahovicek, K., Munteanu, M. G., Gromiha, M. M., Brukner, I., Sanchez, R. and Pongor, S., 1998, Prediction of Bendability and Curvature in Genomic DNA. In *Tenth conversation in biomolecular stereodynamics*, pp. 117-132.
Goodsell, D. S. and Dickerson, R. E., 1994, Bending and Curvature Calculations in B-DNA. *Nucleic Acids Res.* **22**: 5497-503.
Gromiha, M. M., Munteanu, M. G., Gabrielian, A. and Pongor, S., 1996, Anisotropic Elastic Bending Models of DNA. *J. Biol. Phys.* **22**: 227-243.
Gromiha, M. M., Munteanu, M. G., Simon, I. and Pongor, S., 1997, The Role of DNA Bending in Cro Protein-DNA Interactions. *Biophysical Chemistry*, in press.
Han, W., Lindsay, S. M., Dlakic, M. and Harrington, R. E., 1997, Kinked DNA [Letter]. *Nature* **386**: 563.
Langowski, J., Olson, W. K., Pedersen, S. C., Tobias, I., Westcott, T. P. and Yang, Y., 1996, DNA Supercoiling, Localized Bending and Thermal Fluctuations [Letter]. *Trends Biochem. Sci.* **21**: 50.
Langst, G., Schatz, T., Langowski, J. and Grummt, I., 1997, Structural Analysis of Mouse Rdna: Coincidence between Nuclease Hypersensitive Sites, DNA Curvature and Regulatory Elements in the Intergenic Spacer. *Nucleic Acids Res.* **25**: 511-7.
Olson, W. K., 1996, Simulating DNA at Low Resolution. *Curr. Opin. Struct. Biol.* **6**: 242-56.
Olson, W. K., Babcock, M. S., Gorin, A., Liu, G., Marky, N. L., Martino, J. A., Pedersen, S. C., S rinivasan, A. R., Tobias, I., Westcott, T. P. and et, al., 1995, Flexing and Folding Double Helical DNA. *Biophys. Chem.* **55**: 7-29.
Olson, W. K. and Zhurkin, V. B., 1996, Twenty Years of DNA Bending, In *Biological Structure and Dynamics*, (Sarma, R. H. and Sarma, M. H., Eds), Adenine Press, Schenectady, pp. 341-370.
Rippe, K., von Hippel, P. H. and Langowski, J., 1995, Action at a Distance: DNA-Looping and Initiation of Transcription. *Trends Biochem. Sci.* **20**: 500-6.
Sanghani, S. R., Zakrzewska, K., Harvey, S. C. and Lavery, R., 1996, Molecular Modelling of (A4t4nn)N and (T4a4nn)N: Sequence Elements Responsible for Curvature. *Nucleic Acids Res.* **24**: 1632-7.
Satchwell, S. C., Drew, H. R. and Travers, A. A., 1986, Sequence Periodicities in Chicken Nucleosome Core DNA. *J. Mol. Biol.* **191**: 659-75.
Schatz, T. and Langowski, J., 1997, Curvature and Sequence Analysis of Eukaryotic Promoters. *J. Biomol. Struct. Dyn.* **15**: 265-75.
Schellman, J. A., 1974, Flexibility of DNA. *Biopolymers* **13**: 217-26.
Shpigelman, E. S., Trifonov, E. N. and Bolshoy, A., 1993, Curvature: Software for the Analysis of Curved DNA. *Comput. Appl. Biosci.* **9**: 435-40.
Travers, A. A. and Klug, A., 1990, Bending of DNA, In *DNA Topology and Its Biological Effects*, (Cozzarelli, N. R. a. W., J.C., Ed.), Cold Spring Harbor laboratory, Cold Spring Harbor, pp. 57-106.
Trifonov, E. N. and Sussman, J. L., 1980, The Pitch of Chromatin DNA Is Reflected in Its Nucleotide Sequence. *Proc. Natl. Acad. Sci. USA* **77**: 3816-20.
Ulyanov, N. B. and James, T. L., 1995, Statistical Analysis of DNA Duplex Structural Features. *Methods Enzymol.* **261**: 90-120.
Vologodskii, A. V. and Frank-Kamenetskii, M. D., 1992, Modeling Supercoiled DNA. *Methods Enzymol.* **211**: 467-80.

Wheeler, D., 1993, A Gel-Concentration-Independent Retardation Detected in Two Fragments of the Rrnb P1 Promoter of E. Coli Using Transverse Polyacrylamide Pore Gradient Gel Electrophoresis. *Biochem. Biophys. Res. Comm.* **193**: 413-9.

Wu, H. M. and Crothers, D. M., 1984, The Locus of Sequence-Directed and Protein-Induced DNA Bending. *Nature* **308**: 509-13.

Yakushevich, L. V., 1994, Nonlinear DNA Dynamics. *Physica D* **79**: 77-86.

Yang, Y., Tobias, I. and Olson, W.K., 1993, Finite Element Analysis of DNA Supercoiling. *J. Chem. Phys.* **98**: 1673-1686.

Zhang, P., Tobias, I. and Olson, W. K., 1994, Computer Simulation of Protein-Induced Structural Changes in Closed Circular DNA [Published Erratum Appears in J Mol Biol 1995 Sep 1;251(5):721]. *J. Mol. Biol.* **242**: 271-90.

Zhurkin, V. B., Lysov, Y. P. and Ivanov, V. I., 1979, Anisotropic Flexibility of DNA and the Nucleosomal Structure. *Nucleic Acids Res.* **6**: 1081-96.

Zienkiewicz, O. C. a. T., R.L., 1991, *The Finite Element Methods*. McGraw-Hill.

Nonlinear Dynamics in the Binary DNA/RNA Coding Problem

MLADEN MARTINIS
Department of Physics, Theory Division, Ruđer Bošković Institute, P.O.B. 1016, 10001 Zagreb, Croatia

1. INTRODUCTION

The concepts of chaos, nonlinearity and fractal can be applied to a number of properties of proteins. Recently, there has been a rapid accumulation of new information in this area (Elber, 1990, Weigend and Gershenfeld, 1993). However, since proteins are finite systems, although highly inhomogeneous in their structure, their fractal description can only be true in an average or statistical sense. Even if a fractal description is applicable, the question, What new insight does it provide? is sometimes necessary to answer. For proteins this answer is not always clear and straightforward. Nevertheless, deterministic chaos if detected offers a striking explanation for irregular behaviour and anomalies in the protein structure which can not be attributed to the randomness alone.

A protein consists of a polypeptide chain that is made up of residues of amino acids (20 commonly occuring that differ in their side-chains) linked together by peptide bonds. What distinguishes different proteins, other than the number of amino acids, is the sequence of amino acids, referred to as the primary structure of the protein, in the polypeptide chain. From the analysis of many protein structures determined by X-ray crystalography, it is found that the folding of portions of the polypeptide chain often has certain regularities, referred to as elements of secondary structure. The most important secondary structural elements are the α-helices and β-pleated sheets. The α-helix is a compact rod like structure, whereas the β-sheet is an extended structure. Since globular proteins have finite size, the secondary

Supramolecular Structure and Function 7, edited by Pifat-Mrzljak.
Kluwer Academic/Plenum Publishers, New York, 2001.

structural elements (α-helix , β-sheet) are also limited in lenght. Hence the validity of a fractal description as applied to these structures must be taken with care.

In this paper we shall, as one of the tests for nonlinearity, calculate the fractal dimensions for different secondary structure elements of proteins (α-helices and β-sheets). We shall show that fractal dimensions can be used to discriminate an α-helix from a β-sheet. We shall also show how the two-dimensional embedding of the structure can produce an attractor type behavior. The paper is organized in the following way. In Section 1. we introduce the basic mathematical tools necessary for nonlinear analysis of finite protein sequences treated as finite time series. In Section 2. we apply this formalism to DNA/RNA and protein sequences and study a particular aspect of it by finding the fractal dimensions of α-helices and β-sheets using the Hurst exponent approach (Feder, 1989) which measures the smoothness of a sequence.

2. NONLINEAR ANALYSIS OF FINITE SEQUENCES

The most direct link between chaos theory and the real world is the analysis of measured signals/ time series in terms of nonlinear dynamics. In this sense, each nucleotide sequence is represented by a string of suitably chosen numbers $u(n)$, $n = 1,\ldots,N$ which give a novel representation of a sequence that can be further studied using techniques derived from nonlinear analysis (Katz and Schreiber, 1997). Next step is to detect and quantify possibly complicated structures in a sequence which are real and not just random fluctuations. The most convincing argument for the presence of some pattern in the sequence is if it can be used to give an improved prediction of a next term in the sequence. A simple prediction algorithms always assume some deterministic structure in the sequence which can be either linear based on the assumption that the next observation is given approximatelly by a linear combinations of the previous ones plus noise, ξ, i.e.,

$$u(n+1) = a_0 + \sum_{k=1}^{m} a_k u(n-m+k) + \xi(n) \ , \qquad (1)$$

or nonlinear

$$u(n+1) = F[u(n) + \xi(n)] \qquad (2)$$

where the mapping F is usually an unknown nonlinear function. In that case some additional assumption about the properties of F (e.g., continuity) and $\xi(n)$ (e.g., white noise) should be made. One test for nonlinearity is to compare the linear and nonlinear models to see which more accurately predict the next term in the sequence (Casdagli and Weigend, 1993). However, in cases where the considered sequences are not manifestly deterministic and are in addition finite and noisy like in biology, physiology and medicine the usefulness of chaos theory becomes controversal.

It is clear that, a priori, we cannot assume existence of deterministic chaos for any real data that are finite and noisy. The crutial step in resolving this problem is to look for signatures of determinism which can tell us how far we are from the linear case and how close we are to the nonlinear deterministic case. The usual estimators of nonlinearity are:

- nonstationarity
- fractal dimensions
- phase space embeddings
- Lyapunov exponents.

2.1 Nonstationarity

A scientific measurement of any kind is only useful if it is reproducible, at least in principle. Reproducibily is closely connected with the problem of stationarity of sequence measurements. If the process under observation is a probabilistic one, it will be characterised by probability distributions. Stationarity means that they are independent of time translations.

Almost all methods of time series analysis, linear or nonlinear, require some kind of stationarity. For finite sequences the stationarity is not well defined and dependents on the lenght of the sequence. Most of the tests for stationarity are based on the estimate of a certain parameter using different segments of a sequence. If the observed variations are outside the expected statistical fluctuations (standard error), the sequence is considered as nonstationary. In many applications of linear statistics, weak stationarity is assumed which means that the mean and the variance of a sequence are constant. Strict stationarity in addition also requires that the sequence is normally distributed. Therefore, the obvious test for nonstationarity is to consider the length dependece of the mean and the variance of different segments of a sequence:

$$\bar{u}_i(l) = \frac{1}{l}\sum_{k=1}^{l} u(il - l + k) \tag{3}$$

$$\sigma_i^2(l) = \frac{1}{l(l+1)}\sum_{k=1}^{l} (u(il - l + k) - \bar{u}_i(l))^2 \tag{4}$$

where the index i denotes the i-th segment of the sequence $u(n)$ having the length l. For these two quantities the ordering of elements inside the segment is irrelevant and thus they cannot give any information about the evolutin of a system. Such information can be obtained from the sample autocorrelation function

$$c(n) = \frac{1}{N-n}\sum_{k=1}^{N-n}(u(k)-\bar{u})(u(n+k)-\bar{u}) \tag{5}$$

which measures, at fixed lag n, how the points $u(k)$ and $u(n+k)$ are distributed in the plane.

Autocorrelation can be detected visually by examining a regression line scatterplot. Distances from the regression line to the actual values represent error. When no autocorrelation exists, dots representing actual values will be randomly scattered around the full length of the series. The value of the autocorrelation function at lag n=0 is the variance σ^2 of the sequence.

The autocorrelation function $C(n)$ may be used to detect deterministic components masked in a random background because deterministic data usually have long range correlations in which $C(n)\sim\mathbf{n}^{-\beta}$ for large n, while random data have short-range correlations and $C(n)$ decays exponentially for large n. Observation of long-range correlations in a sequence may be a first sign for underlying nonlinear dynamics.

Another useful characteristic of nonstationarity is the behaviour of the low-frequency part of the power spectrum which is defined as the Fourier transform of the autocorrelation function $C(n)$. An equivalent definition of the power spectrum is the squared modulus of the Fourier transform of a sequence $u(n)$:

$$P(f) = \left|\frac{1}{\sqrt{N}}\sum_{n=1}^{N}(u(n)e^{2\pi ifn}\right|^2 \tag{6}$$

where the frequencies in physical units are $f\equiv f_k=k/N$, k=-N/2,...,N/2. The relation between the power spectrum $P(f)$ and the autocorrelation function $C(n)$ is known as the Wiener-Khinchin theorem. The sequencies with the power spectra at low frequences proportional to $|f|^{-\alpha}$ with positive α are considered nonstationary. This trend usually characterizes the most biological signals. White noise is defined by a flat spectrum; it has a spectral exponent α= 0. When white noise is summed up over n, the result is a random walk process with a spectral exponent α=2, known as a brown noise. The intermediate case with α= 1, known as a pink noise, is encountered in a wide variety of physical, chemical, and biological systems. It implies that all lenght-scales in a sequence, over which $1/f$ behaviour holds, are equally important. A power spectrum with $\alpha > 2$ is known as a black noise; it governs the phenomena connected with natural and unnatural catastrophes which often come in clusters.

For fractal sequences to be discussed below there is a simple relation between the exponent α and the fractal dimension D of the sequence given by $\alpha = 5 - 2D$.

2.2 Fractal dimensions

The recent realization that nonlinear dynamical systems can display deterministic chaos has had a strong influence on the researchers to interpret complex physiological data in terms of chaos. A number of techniques based on concepts from nonlinear dynamics have been proposed (Grassberger *et al.*, 1991). The fractal dimension (Farmer, 1982; Grassberger and Procaccia, 1983; Grassberger, 1983) as one of them gives a statistical measure of the geometry of the clouds of points. In deterministic chaotic systems the dimension is not always an integer number but fractional. Therefore, fractals are disordered systems whose disorder can be described in terms of noninteger dimensions. The focal point of fractal geometry is the repetition of disorder at all length scales (scale symmetry), or simply, the repetition of structure within structure. There are various ways and levels of precision of defining and calculating the fractal dimensions. We shall present here the calculation of the fractal dimension of an irregular sequence by means of the Hurst's rescaled range analysis exponent (Feder, 1989).

For a given sequence $u(n)$, $n = 1,\ldots,N$ the rescaled range is defined as the range $R(n)$ of the sum of the deviations of data from the mean divided by the sample standard deviation $S(n)$. The relevant quantities are defined in the following way. The sample mean over the length n is

$$\bar{u}(n) = \frac{1}{n}\sum_{k=1}^{n} u(k) \tag{7}$$

it is used to construct the accumulated departure $X(l,n)$ of $u(k)$ from the mean $\bar{u}(n)$,

$$X(l,n) = \sum_{k=1}^{l} \left[u(k) - \bar{u}_i(n)\right] . \tag{8}$$

The range $R(n)$ is obtained as the difference between the maximum and the minimum of $X(l,n)$ for fixed n,

$$R(n) = \max_{l} X(l,n) - \min_{l}(l,n) . \tag{9}$$

The standard deviation $S(n)$ is estimated from

$$S(n) = \left(\frac{1}{n} \sum_{k=1}^{n} [\bar{u}(n)]^2 \right)^{\frac{1}{2}} \tag{10}$$

The rescaled range is then the dimensionless ratio *R/S*. The Hurst exponent *H* is defined in the range of n where the Hurst law $R/S \sim n^H$ is valid. In general, *H* is not a constant but a n-dependent quantity which can be estimated using the log(*R/S*) versus log(*n*) plot. For these reasons the *R/S* analysis is considered as robust. The Hurst exponent is used to estimate the smoothness of an irregular sequence. Thus, if $H = 0.5$, the behaviour of the sequence is similar to a random walk;
if $H< 0.5$, the sequence is very spiky and is dominated by short-range fluctuations;
if $H> 0.5$, the sequence is less spiky and is dominated by long-range fluctuation.

There is a simple relation between *H* and the fractal dimension *D*:

$$D = 2 - H \tag{11}$$

H is related to the "1/*f*" spectral slope $\alpha = 2H + 1$ and can also be estimaded by means of the Detrendet Fluctuation Analysis (DFA) (Peng 1995) used for quantifying the correlation properties in a nonstationary sequence. DFA is based on the computation of a scaling exponent "*d*" by means of a modified root mean square analysis of a random walk: $d = H + 1$.

2.3 Phase space embeddings

When analysing a finite sequence *u(n)*, $n = 1,....,N$, we will almost always have only incomplete information about the underlying dynamical system which may live in a high-dimensional state space. Therefore, in order to construct a dynamical model for a given sequence of data it is necessary first to reconstruct a state space. One technique to do so is, for example, the phase plane embedding or phase portrait of a sequence, which involves plotting $u(n + \tau)$ vs. $u(n)$ with time lag τ, giving a trajectory in the two-dimensional phase space. Embeddings of a sequence in higher dimensional phase space rest on the assumption that a genuine trajectory $\mathbf{x}(t)$ of a dynamical system is a state vector in a *d*-dimensional space. The delay embedding theorem by Takens (1981) states that a τ-delayed reconstructed state vector

$$\mathbf{u}(n) = (u(n), u(n-\tau),..., u(n-E\tau+\tau)) \tag{12}$$

in E-dimensional embedding phase space will, in the absence of noise, approximate dynamical system provided $E \geq 2d + 1$; in practise $E > 2d_{Box}$, where d_{Box} is the box counting fractal dimension of a sequence (Sauer, 1991). For a deterministic dynamical system a phase-space description is an optimal way of studying its dynamical and geometrical properties. The problem of finding a good embedding is to estimate the embedding dimension E and to find the correct time lag τ (Katz and Schreiber, 1997).

2.4 Lyapunov exponents

So far we have discussed predictive criteria for finding nonlinear dynamics in a sequence of data. A parameter for diagnosing whether or not a dynamical system is chaotic are the Lyapunov exponents which measure the average local exponential growth rate of infinitesimal initial errors, that is, of neighboring trajectories in phase-space embeddings (Wolf, 1985).

If $\mathbf{u}(n_1)$ and $\mathbf{u}(n_2)$ are two points in state space with the Euclidean distance $\|\mathbf{u}(n_1) - \mathbf{u}(n_2)\| = d_0 \ll 1$ then the distance after a time δn between the two trajectories will be $\|\mathbf{u}(n_1 + \delta n) - \mathbf{u}(n_2 + \delta n)\| = d_{\delta n}$. Then the Lyapunov exponent λ is defined by

$$d_{\delta n} \cong d_0 e^{\lambda \delta n} , \tag{13}$$

with $d_{\delta n} \ll 1$ and $\delta n \ll 1$. If a system is known to be deterministic, a positive Lyapunov exponent can be taken as a definition of a chaotic system. Critical issues in the reliability of the estimation of Lyapunov exponents are the influence of noise.

3. FRACTAL DIMENSIONS OF DNA SEQUENCES IN ALL-α AND ALL-β PROTEINS

In this Section we report on the finding the fractal dimensions of DNA sequences appearing in the secondary strucures of proteins (Martinis, *et al.*). We have analysed 12 α-helices and 12 β-sheets using the Hurst exponent approach which measures the smoothness of a given sequence. Since the Hurst exponent H is related to the fractal dimension $D = 2 - H$ of the sequence, to its "$1/f$" spectral slope $\alpha = 2H + 1$ and to the DFA slope $d = H + 1$ it is a suitable parameter for studying correlation properties in DNA sequences.

A DNA sequence consists of four kinds of bases: A, C, G and T/U. The bases can be divided in classes: purine ($R = A, G$), pyrimidine ($Y = C, T/U$), strong H-bonds ($S = G, C$) and weak H-bonds ($W = A, T/U$). In order to analyse symbolic DNA sequences the assignment of numbers to the nucleotides *A*, *C*, *G* and *T*/*U* is necessary. Taking into account the classes we choose the following binary assignment:

$$Y \to 0\,,\ S \to 1 \tag{14}$$

$$R \to 1\,, W \to 0\,. \tag{15}$$

This assignment maps the simbolic nucleotide sequence on a sequence $u(n)$, $n = 1,\ldots, N$(length of the sequence) where $u(n)$ can take values 0 or 1 only. Similar approach, based on the purine-pyrimidine classification sheme, has been proposed by Peng *et al.*(1992). Our analysis has shown that α-helices and β-sheets have different fractal dimensions, $D_\alpha < D_\beta$ with $D_\alpha \approx$ 1.41.

We have also studied the behaviour of sequences in 2-dimensional phase-space embeddings by first transforming the digital sequence $u(n)$ into an analoge signal by means of the scaled sinc(t) function:

$$A(t) = \sum_{k=1}^{N} u(k) \sin c\left(\frac{t}{2} - k\right)\,, \tag{16}$$

$$\sin c(t) = \frac{\sin(\pi t)}{\pi t}\,. \tag{17}$$

The phase portrait $A(t + 1)$ vs. $A(t)$ shows the attractor type behaviour.

4. CONCLUSION

In this paper we have discussed several simple methods which are useful for analysing and estimating complex structures in DNA sequences. We have argued using the Hurst exponent approach that fractal dimensions of DNA sequences in all-α and all-β proteins are different. Observation of scale dependent fractal dimension (breaks in the slopes of log-log plots of R/S) may indicate a transition from one pattern of variation at one scale to another pattern at a larger scale. There are probably several contributing causes to these observed variations in D. Although the DNA sequences exibit statistical characteristic which might be explained by nonlinear statistical methods, no definite sign of chaos has been evidenced yet.

ACKNOWLEDGMENT

This work was supported by the Ministry of Science of the Republic of Croatia under Contract No. 980104.

REFERENCES

Casdagli, M., and Weigend, A.S., 1993, Exploring Continuum between Deterministic and Stochastic Modeling. In *Proc. of Santa Fe Institute Studies in the Sciences of Complexity*, Addison-Wesley, Reading, MA, Vol. XV.

Elber, R., 1990, Fractal Analysis of Proteins. In *The Fractal Approach to Heterogeneous Chemistry*, (D. Anvir, ed.) John Willey & Sons, NY.

Farmer, J.D., 1982, Chaotic attractors of an infinite-dimensional chaotic system. *Physica* **4D**: 366-393.

Feder, J., 1989, *Fractals*, Plenum Press, NY.

Grassberger, P., and Procaccia, I., 1983, Measuring the strangenes of strange attractor. *Physica* **9D**: 189-208.

Grassberger, P., 1983, Generalized dimension of strange attractors. *Phys. Lett.* **A97**: 227-230.

Grassberger, P., Schreiber, T., and Schaffrath, C., 1991, Nonlinear time sequence analysis. *Intl. J. Bif. & Chaos* **1**: 512-520.

Katz, H., and Schreiber, T., 1997, *Nonlinear Time Series Analysis*, Cambridge University Press, Cambridge, UK.

Martinis, M., Štambuk, N., and Črnugelj, J., Fractal dimensions of α-helices and β-sheets, in preparation.

Packard, N.H., Crutchfeld, J.P., Farmer, J.D., and Show, R.S., 1980, Geometry from a time series. *Phys. Rev. Lett.* **45**: 712-716.

Peng, C.K., Havlin, S., Stanley, H.E., and Goldberger, A.L., 1995, Quantification of scaling exponents and crossover phenomena in nonstationary heartbeat time series. *Chaos* **5**: 82.

Peng. C.K., Buldyrev, S.V., Goldberger, A.L., Havlin, S., Simon, M., and Stanley, H.E., 1992, Long-range correlations in nucleotide sequences. *Nature* **356**: 168-170.

Proceedings of Santa Fe Institute Studies in the Sciences of Complexity, 1993, (A. S. Weigend and N. A. Gershenfeld, eds.) Addison-Wesley, Reading, MA, Vol. XV.

Ruelle, D., 1990, Deterministic chaos: The Science and the Fiction. *Proc. Roy. Soc. London Ser.* **A427**: 241-248.

Sauer, T., Yorke, J.A., and Casdagli, M., 1991, Embedology. *J. Stat. Phys.* **65**: 579-616.

Takens, F., 1981, Detecting Strange Attractors in Turbulence. In *Dynamical Systems and Turbulence*, (D. Rand and L.-S. Young, eds.), Springer-Verlag, Berlin.

Wolf, A., Swift, J.B., Swinney, H.L., Vastano, J.A., 1985, Determining Lyapunov Exponents from a Time Series. *Physica* **16D**: 285-317.

High Density DNA and Protein Arrays, Generation and Applications

DOLORES J. CAHILL
Max-Planck-Institute of Molecular Genetics, Ihnestrasse 73, D-14195 Berlin, Germany and PROTAGEN, Im Lottental 36, 44801 Bochum, Germany

1. INTRODUCTION

High-density DNA and protein arrays, also known as microarrays or chips, are miniaturised devices, which comprise small flat surfaces, onto which are ordered arrangements of individual samples are positioned, enabling their parallel analysis, which can be used for biological experimentation. High density array technologies allow the use of smaller sample volumes, more efficient analyses and higher throughput. This review outlines recent developments in the generation of high density DNA and protein arrays and their potential applications in the areas such as diagnostics and proteomics.

2. DNA ARRAYS

For DNA arrays, PCR techniques are well established and play a central role in large-scale genome analysis. Chip production is highly automated, using either pin-based or micro-dispensing liquid handling robots. High density DNA arrays can be generated in two ways, one method involves needle or pin spotting, where the liquid containing the DNA fragment is delivered through adhesion to stainless steel pins. The second method utilises piezo-ink-jet technology, where cDNAs, for example, are transferred without touching the surface (Eickhoff *et al.*, 1998). The samples to be arrayed are generally held in microtitre plates. The 96-well microtitre plate is still the most widely used in immunoassays, whereas 384-well plates also used for many assays and are now standard for storage and handling of clone libraries. 1,536-well plates are also available, and etched glass or silicon wafers make substrates for so-called 'nanoplates' (e.g. 9,600-wells; 100-1000nl well volumes) and other microarray formats which far exceed the density potential of the standard microtitre plate have been developed. (Walter *et al.*, 2000). If large numbers of samples are to be analysed for their interaction with the same ligand, micro (spot) arrays on planar surfaces are the format of choice. Those surfaces are either filter membranes (e.g. nitrocellulose, nylon or polyvinylidene difluoride) or glass slides coated with various reagents (e.g. poly-L-lysine or polyacrylamide).

3. PROTEIN ARRAYS

Since proteins translate genomic sequence information into function thereby enabling biological processes, a full understanding of the expression profile of a tissue or organism on both the genomic and proteomic levels requires the screening of many samples in parallel, as rapidly as possible. However, as proteins are more complex biomolecules and diverse biomolecules than nucleic acids, this greatly complicates the development of unified platforms for the screening of many analytes. Until recently, it had not been possible to analyse proteins using the same high-density array, automated approach described for the DNA arrays. We have applied automation technologies to the high-throughput, large-scale analysis of proteins, by generating cDNA expression libraries, high-density protein arrays (Büssow *et al.*, 1998), and micro-arrays (Lueking *et al.*, 1999). Protein arrays can be generated using two methods, either by arraying the clones on a membrane and after overnight growth, and subsequent induction of protein expression on the membrane or filter (Büssow *et al.*, 1998) or by expression and purification of each proteins from the library in microtitre plates, and arraying the purified proteins onto filters or chips (Lueking *et al.*, 1999).

The first method involved using robotic technology, to array bacterial colonies from a human foetal brain cDNA expression library (hEx1), which were subsequently gridded onto PVDF filters, producing high density colony protein filters. *In situ* expression of recombinant fusion proteins was induced and detected using an antibody against a His-tag-containing epitope (Büssow *et al.*, 1998). Using this approach, the genes from a cDNA library can be studied on the DNA and protein levels simultaneously, and provide sources of recombinant genes and proteins to make DNA and protein chips. This approach has been extended to produce the second type of protein array which involves the automated spotting of purified protein micro-arrays from liquid expression cultures using a transfer stamp mounted onto a flat-bed spotting robot (Lueking *et al.*, 1999), which showed that such protein micro-arrays provide the means for very sensitive gene expression and antibody specificity screening in high-throughput. Using this transfer stamp, which had a diameter of 150μm instead of 450μm, 4800 samples can be placed onto a microscopic slide and simultaneously screened, applying minimal amounts of reagents. Sharp and well localised signals allowed the detection of 250 attomoles or 10pg of a spotted test protein (GAPDH). The protein expression clones of the cDNA library were reliably detected, with a low rate (11%) of false positive clones expressing proteins in incorrect reading frames. These improved protein micro-arrays allow very sensitive protein expression and antibody specificity screening. Such approaches make protein products encoded by cDNA clones available for further analysis and use, such as the generation of protein arrays or of MPIs. This requires a highly parallel approach to protein expression analysis, including the simultaneous expression of large numbers of cDNA clones in an appropriate vector system and high-speed arraying of protein product (Cahill *et al.*, 2000).

Further recent developments in protein arrays include the generation of low-density protein arrays on filter membranes, such as the universal protein array system (UPA), which is based on the 96-well microtitre plate format (Ge, 2000). Protein micro-arrays have also been printed, again in the microtitre well format, on an optically flat glass plate containing 96 wells formed by an enclosing hydrophobic Teflon mask (Mendoza *et al.*, 1999). Inside the wells, arrays of 144 (4 x 36) elements each, were spotted using a 36-capillary-based print head attached to a precise, high-speed, X-Y-Z robot. Standard ELISA techniques and a scanning CCD detector were employed for imaging of arrayed antigens.

Other approaches to protein microarrays have been reported using either photolithography of silane monolayers (Mooney *et al.*, 1996) or gold (Jones *et al.*, 1998), combining microwells with microsphere sensors (Michael *et al.*, 1998) or ink-jetting onto polystyrene film (Ekins, 1998). Those advances are focused on the fabrication of miniaturised immunoassay formats by patterning of single proteins (e.g., BSA, avidin or monoclonal antibodies).

Surface plasmon resonance (BIAcore) has contributed greatly to interaction studies in a chip format. SELDI and BIA-MS technology combine affinity capture and high resolution mass analysis are combined into novel and efficient means for the characterisation of proteins and their interaction partners, and has applications such as epitope mapping and receptor-ligand or protein-drug interaction studies. For capturing and analysis of specifically labelled proteins, ligand-coated surfaces are available, like the SELDI ProteinChip (Ciphergen Biosystems Inc., Palo

Alto, CA, USA, www.ciphergen.com), XNA on Gold (INTERACTIVA Biotechologie GmbH, Ulm, Germany, www.interactiva.de) and various BIAcore chips (Biacore AB, Uppsala, Sweden, www.biacore.com).

4. ANTIBODY, TISSUE, AND OTHER ARRAY FORMATS

A method for fabricating antibody arrays has been reported which uses a micro-moulded hydrogel 'stamper' and an aminosilylated receiving surface (Martin *et al.*, 1998). The stamper deposits protein as a submonolayer, as shown by I-125 labelling and atomic force microscopy, whilst antibody activity was retained.

An array immunosensor has been developed which allowed the simultaneous detection of clinical analytes (Rowe *et al.*, 1999). This was also in a microtitre plate format and had capture antibodies and analytes arrayed onto microscopic slides using flow chambers in a cross-wise fashion. Detection was again via fluorescent labels and CCD-based optical readout. Although still at a low-density stage (6 x 6 pattern), the technique has high-throughput potential as it involves automated image analysis and micro-fluidics which is already becoming one of the future formats for enzyme activity and other assays (Cohen *et al.*, 1999).

Tissue micro-arrays have recently been developed for high-throughput molecular profiling of tumor specimens (Kononen *et al.*, 1998), which involved a robot punching cylinders (0.6 mm wide, 3-4 mm high) from 1000 individual tumour biopsies embedded in paraffin and to array them in a 45 x 20 mm paraffin block (commercially available from Beecher Instruments, Silver Spring, MD, USA, www.beecherinstruments.com). On serial sections, tumours are then analysed in parallel by immunohistochemistry, fluorescence in situ hybridisation (FISH) and RNA/RNA in situ hybridisation. This system allowed the microscopic scanning of an immunohistochemistry array slide containing 645 specimens in less than two hours. It should be of great help for the simultaneous analysis of tumors from many different patients at different stages of disease to establish the therapeutic importance of new candidate marker genes more rapidly (Theillet, 1998).

To manufacture three-dimensional arrays on a flat surface, a gel photo- or persulfate-induced co-polymerisation technique has been developed to produce oligonucleotide, DNA and protein microchips on polyacrylamide gel pads from 10 x 10 to 100 x 100 µm, separated by a hydrophobic glass surface (Guschin *et al.*, 1997). The three-dimensional polyacrylamide gel provides a more than 100 times greater capacity for immobilisation than does a two-dimensional glass support, thus increasing the sensitivity of measurements considerably (Parinov *et al.*, 1996).

5. FUTURE DEVELOPMENTS

Further developments required, include increasing the output from DNA and protein chips, which would involve increasing the quality of the data output, high speed arraying, high throughput hybridisation devices with on-line detection, plus integrating image and data analysis tools.

Further developments have been made in further miniaturising arraying and micro-arraying technologies such as the construction and performance of a fully automated multi-capillary electrophoresis system for the analysis of fluorescently labelled biomolecules such as DNA (Behr *et al.*, 1999). In this device, a special detection system allows the simultaneous spectral analysis of all 96 capillaries, which has no moving parts, is highly robust, and is fully compatible with existing systems. The device can automatically process up to 40 MTPs (96 and 384-well MTPs) that is up to 15,000 samples simultaneously without human interference, before it has to be re-loaded.

Other application of the micro-dispensing technology is the preparation of high-density arrayed probes on target plates for MALDI-MS. Mass spectrometry allows an enormous speed for high-throughput applications in DNA and protein analysis. However, some drawbacks are limiting the use of MALDI-MS in biotechnology. For proteome projects, the protein samples have to be highly purified from salts and detergents, for example SDS (sodium dodecylsulphate) from SDS-PAGE gels or staining reagents, increase drastically the background during the measurement. A high-throughput MALDI-TOF mass spectrometry technique has been developed for robust, rapid and highly accurate quality control of proteins (Büssow *et al.*, 2000). Small deviations from expected masses, resulting from chemical modification or loss of single amino acids, can be efficiently detected. We have proposed (Cahill *et al.*, 2000) and shown (Schmidt *et al.*, 2001) that the method, technical feasibility and applications of characterising a protein by a minimal set of structural information generated by mass spectrometry, called 'minimal protein identifier' (MPI). MPIs can be generated from proteins from 2-D gels and recombinant proteins, as obtained from a cDNA library, which can be used to identify and compare proteins from these sources, allowing a connection between 2-D gels, protein and DNA arrays. The development of such automated approaches with very high-throughput is required for such protein analysis, since large numbers of proteins are involved in the homeostasis of even the simplest organisms (Uetz *et al.*, 2000).

6. APPLICATIONS

The main applications of DNA arrays are for gene expression profiling or polymorphism analysis.

Protein arrays enable high-throughput screening for gene expression and molecular interactions. One of the main advantage of the use of cDNA libraries as the source of proteins is that this approach achieves the large-scale systematic provision of recombinant proteins for functional studies by making and arraying cDNA expression libraries and by allowing the direct connection from DNA sequence information on individual clones to protein products and back again on a whole genome level. This makes translated gene products directly amenable to high-throughput experimentation and generates a direct link between protein expression and DNA sequence data. Since protein micro-arrays are a useful tool for connecting gene expression analysis and molecular binding studies on a whole-genome level, if differentially expressed genes are identified using cDNA micro-arrays, the same clones can be analysed simultaneously for protein expression in different cellular systems or by *in vitro* transcription/translation. On identical protein micro-arrays, expression clones can be screened for binding to other proteins (e.g., antibodies) or to diverse molecules from nucleic acids to small-molecule ligands. This versatility should make protein micro-arrays a multi-purpose tool for diagnostic use. We have recently described their use as powerful tools for high-throughput ligand-receptor interaction studies, diagnosis and antibody specificity characterisation (Walter *et al.*, 2000). For example, the low-density universal protein array system (UPA) has been used for studying interactions with protein, DNA, RNA and small chemical ligand probes (Ge, 2000). Testing interactions of the human protein p52 with 48 purified proteins spotted onto a nitrocellulose membrane, it was possible to distinguish high-affinity protein-protein interactions by washing the membrane with high salt concentrations (500-1,000 mM KCl). High-density protein filters of the hEx1 library were screened with antibodies against the human proteins GAPDH and HSP90 alpha, and positives were confirmed bycDNA probing and sequencing(Büssow *et al.*, 1998). hEx1 protein filters have now been used by a number of groups to detect expression clones using antibodies, single-chain Fv fragments or other protein-protein or nucleic acid-protein interaction screening (Walter *et al.*, 2000).

Protein array technologies also have further applications such as the serum screening, for example from patients with auto-immune antibodies. (Joos *et al.*, 2000). It is envisaged that such screenings will generate novel diagnostic targets, a method of characterising the particular disease based on the antibody profile of the patient, and as a prognostic tools.

7. CONCLUSIONS

DNA, protein and antibody arrays appear as new and versatile tools in functional genomics and proteomics. An ever increasing variety of array formats becomes available, e.g. microtitre plates, patterned arrays, three-dimensional pads, flat surface spot arrays or micro-fluidic chips. DNA and proteins can be arrayed onto different surfaces, e.g. membrane filters, polystyrene film, glass, silane or gold. It is envisaged that any ligand-binding assay that relies on detecting the product of an immobilised capture molecule and its binding partner can be perform in a biochip format.

REFERENCES

Behr, S., Matzig, M., Levin, A., Eickhoff, H., and Heller, C., 1999, A fully automated multicapillary electrophoresis device for DNA analysis. *Electrophoresis* **20**: 1492-507.

Büssow, K., Nordhoff, E., Lübbert, C., Lehrach, H., and Walter, G., 2000, A human cDNA library for high-throughput protein expression screening. *Genomics* **65**: 1-8.

Büssow, K., Cahill, D., Nietfeld, W., Bancroft, D., Scherzinger, E., Lehrach, H., and Walter, G., 1998, A method for global protein expression and antibody screening on high - density filters of an arrayed cDNA library. *Nucleic Acids Research* **26**: 5007-5008.

Cahill, D. J., Nordhoff, E., O'Brien, J., Klose, J., Eickhoff, H., and Lehrach, H., 2000, Bridging Genomics and Proteomics. In *Proteomics. From protein sequence to function.* (Pennington, S. R., and Dunn, M. J., Eds.), BIOS Scientific Publishers Ltd., pp. 1-17.

Cohen, C. B., Chin, D. E., Jeong, S., and Nikiforov, T. T., 1999, A microchip-based enzyme assay for protein kinase A. *Anal. Biochem.* **273**: 89-97.

Eickhoff, H., Cahill, D., Nordhoff, E., Heise, V., and Lehrach, H., 1998, DNA and protein expression monitoring on arrays, In *Genetic Diagnosis in Medicine*. (Garcia-Foncillas, J., Ed.), Slide Print, S.L., Pamplona, pp. 503-511.

Ekins, R. P., 1998, Ligand assays: from electrophoresis to miniaturized microarrays. *Clinical Chemistry* **44**: 2015-2030.

Ge, H., 2000, UPA, a universal protein array system for quantitative detection of protein-protein, protein-DNA, protein-RNA and protein-ligand interactions. *Nucleic Acids Research* **28**: e3.

Guschin, D., Yershov, G., Zaslavsky, A., Gemmell, A., Shick, V., Proudnikov, D., Arenkov, P., and Mirzabekov, A., 1997, Manual manufacturing of oligonucleotide, DNA, and protein microchips. *Anal. Biochem.* **250**: 203-211.

Lueking, A., Horn, M., Eickhoff, H., Bussow, K., Lehrach, H., and Walter, G., 1999, Protein microarrays for gene expression and antibody screening. *Anal. Biochem.* **270**: 103-111.

Martin, B. D., Gaber, B. P., Patterson, C. H., and Turner, D. C., 1998, Direct protein microarray fabrication using a hydrogel "stamper". *Langmuir* **14**: 3971-3975.

Mendoza, L. G., McQuary , P., Mongan, A., Gangadharan, R., Brignac, S., and Eggers, M., 1999, High throughput microarray-based enzyme-linked immunosorbent assay (ELISA). *Biotechniques* **27**: 778-788.

Michael, K. L., Taylor, L. C., Schultz, S. L., and Walt, D. R., 1998, Randomly ordered addressable high-density optical sensor arrays. *Anal. Chem.* **70**: 1242-1248.

Mooney, J. F., Hunt, A. J., McIntosh, J. R., Liberko, C. A., Walba, D. M., and Rogers, C. T.,

1996, Patterning of functional antibodies and other proteins by photolithography of silane monolayers. *Proc. Natl. Acad. Sci. USA* **93**: 12287-12291.

Joos, T. A., Schrenk, M., Höpfl, P., Kröger, K., Chowdhury, U., Stoll, D., Schörner, D., Dürr, M., Herick, K., Rupp, S., Sohn, K., and Hämmerle, H., 2000, A microarray enzyme-based immunosorbent assay for autoimmune diagnostics. *Electrophoresis* **21**: 2641-2650.

Jones, V. W., Kenseth, J. R., Porter, M. D., Mosher, C. L., and Henderson, E., 1998, Microminiaturized immunoassays using atomic force microscopy and compositionally patterned antigen arrays. *Anal. Chem.* **70**: 1233-1241.

Kononen, J., Bubendorf, L., Kallioniemi, A., Barlund, M., Schraml, P., Leighton, S., Torhorst, J., Mihatsch, M. J., Sauter, G., and Kallioniemi, O. P., 1998,. Tissue microarrays for high-throughput molecular profiling of tumor specimens. *Nat. Med.* **4**: 844-847.

Parinov, S., Barsky, V., Yershov, G., Kirillov, E., Timofeev, E., Belgovskiy, A., and Mirzabekov, A., 1996, DNA sequencing by hybridization to microchip octa-and decanucleotides extended by stacked pentanucleotides. *Nucleic Acids Research* **24**: 2998-3004.

Rowe, C. A., Scruggs, S. B., Feldstein, M. J., Golden, J. P., and Ligler, F. S., 1999, An array immunosensor for simultaneous detection of clinical analytes. *Anal. Chem.* **71**: 433-439.

Schmidt, F., Lueking, A., Nordhoff, E., Gobom, J., Klose, J., Seitz, H., Eickhoff, H., Lehrach, H., and Cahill, D. J., 2001, Generation of minimal protein identifiers of proteins from 2D gels and recombinant proteins. *Electrophoresis*, in press.

Theillet, C., 1998, Full speed ahead for tumor screening. *Nat. Med.* **4**: 767-768.

Uetz, P., Giot, L., Cagney, G., Mansfield, T. A., Judson, R. S., Knight, J. R., Lockshon, D., Narayan, V., Srinivasan, M., Pochart, P., Qureshi-Emili, A., Li, Y., Godwin, B., Conover, D., Kalbfleisch, T., Vijayadamodar, G., Yang, M., Johnston, M., Fields, S., and Rothberg, J. M., 2000, A comprehensive analysis of protein-protein interactions in *Saccharomyces cerevisiae*. *Nature* **403**: 623-627.

Walter, G., Büssow, K., Cahill, D., Lueking, A., and Lehrach, H., 2000, Protein arrays for gene expression and molecular interaction screening. *Current Opinion in Microbiology* **3**: 298-302.

Mechanisms of Synaptic Plasticity in the Brain

KREŠIMIR KRNJEVIĆ
Anaesthesia Research Department, McGill Iniversity, Montréal QC, H3G 1Y6, Canada

1. INTRODUCTION

Synapses are the main sites of interactions between nerve cells (neurons). The Oxford English Dictionary defines "plasticity" as "aptness to be molded". In current neuroscientific discourse, synaptic plasticity is the capacity of synaptic connections to become stronger or weaker as a result of previous activity. Two broad types of alterations in synaptic function include: on the one hand, those involved in growth, maturity or senescence; and on the other, those mediating the long-term changes which are widely believed to be the basis of learning and memory. My topic deals pre-eminently with the type of synaptic plasticity, which allows mature synapses to undergo "long-term" changes in potency. Because it may be the cellular basis for learning, such plasticy is now of tremendous interest for the neuroscience community.

Until a few decades ago, learning and memory were viewed as the concern of psychologists and philosophers, and not amenable to neurophysiological investigation. Most psychologists treated these as psychic processes that might or might not involve specific brain activities. Even such an outstanding investigator of learning as I.P. Pavlov -- who was trained as a physiologist -- was content to ignore basic mechanisms operating at the cellular and synaptic level (Pavlov, 1927). By the middle of the 20th century, however, several authors were making serious attempts to bridge the gap between studies of learning and basic neurophysiology. The general consensus was that, bearing in mind the high speed of learning, the most likely mechanism was increased potency of synaptic connections after repeated activation (Konorski, 1948, Hebb, 1949, Eccles, 1953). Albeit both lucid and insightful (pointing in the right direction as things turned out), these authors fully realized that their speculations were not based on any solid experimental evidence.

Supramolecular Structure and Function 7, edited by Pifat-Mrzljak.
Kluwer Academic/Plenum Publishers, New York, 2001.

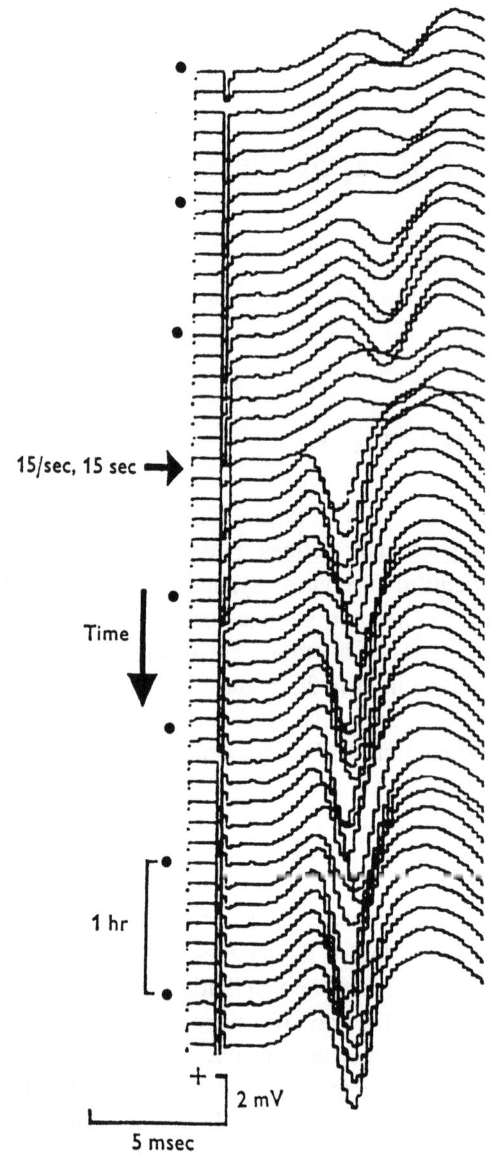

Figure 1. Long-lasting potentiation of synaptic responses in dentate gyrus of unanaesthetized rabbit. Sequence of traces (from above down) shows averages of 16 consecutive responses to stimulation of perforant path at 0.5/min (at constant intensity throughout), recorded over several hours from freely-moving animal. Note fluctuations of control traces before high frequency train (at horizontal arrow) and large increase in amplitude and reduction in latency afterwards (from Fig. 4 in Bliss and Gardner-Medwin, 1973). (A diagram illustrating the hippocampus and the placement of stimulating and recording electrodes is in right half of Fig. 4A).

Though it was known that transmission at the nerve-muscle junction, autonomic ganglia and even the spinal cord can be facilitated by repeated stimulation -- for example, high-frequency stimulation results in "post-tetanic potentiation" (PTP). But such changes last at most only a few minutes. They could hardly be mechanisms of memories that persist for hours, days and even much longer. The situation did not change for another two decades, until Bliss and Lømo's (1973) epoch-making discovery that high frequency ("tetanic") stimulation can induce a potentiation of hippocampal synaptic responses lasting many hours (long-term potentiation or LTP) -- even in freely-moving non-anaesthetized animals (Fig. 1). Clearly, this was the kind of change postulated by Konorski (1948), Hebb (1949) and Eccles (1953).

Particularly significant was the location of LTP. The hippocampus, together with adjoining temporal cortex, had for some time been suspected to be involved in mechanisms of memory, notably because surgical removal (for intractable epilepsy) produced a major disturbance of memory (Scoville and Milner, 1957). Bliss and Lømo's (1973) finding (generally known as long-term potentiation or LTP) was confirmed in numerous later experiments, especially in various regions of the hippocampus, but also in some other brain structures (Bliss and Collingridge, 1993).

Unlike the multilayered neocortex, the dentate gyrus and the neighbouring hippocampus consist of only a single layer of pyramidal cells (Fig. 4A): the simpler arrangement of neurons and connections within this region greatly facilitates electrophysiological analyses. For this reason, the hippocampus has been favoured for studies of long-term changes in synaptic potency -- especially since the advent of brain slices. In thin slices, maintained in well-oxygenated saline, neuronal and synaptic function is preserved for many hours. Because slices are mechanically also very stable, synaptic potentials (excitatory post-synaptic potentials or EPSPs) can be recorded for long periods: with a relatively large extracellular microelectrode -- which "sees" global responses of populations of cells ("field" responses) (eg. traces in Figs. 1-3 and 8) -- or very fine electrodes inserted in a single cell (lower inset traces in Fig. 3). In such *in vitro* experiments, the investigator has full control over the ionic composition of the external milieu, and even that of the cell interior (when using somewhat larger "patch" electrodes). With the voltage-clamp technique, membrane currents (including excitatory post-synaptic currents or EPSCs, Figs. 6 and 7) and specific changes in membrane conductances (voltage-dependent or ligand-gated) can be analysed in great detail.

By the early 1980's, it was clear that, under the right conditions -- tetanic stimulation (typically 100 Hz for 1 s) had to be of sufficient intensity to induce substantial depolarization of the post-synaptic cells -- LTP was a highly reproducible phenomenon at hippocampal synapses. By contrast, prolonged low frequency stimulation (≈1 Hz) tended to have an opposite effect: either lasting *depression* (LTD) of transmission (Linden and Connor, 1995) or "depotentiation" of an existing LTP. Like LTP, LTD selectively affects the synapses that have been stimulated repetitively; though also observed in the hippocampus (Fig. 2), it is more prominent in neocortex and especially the cerebellar cortex.

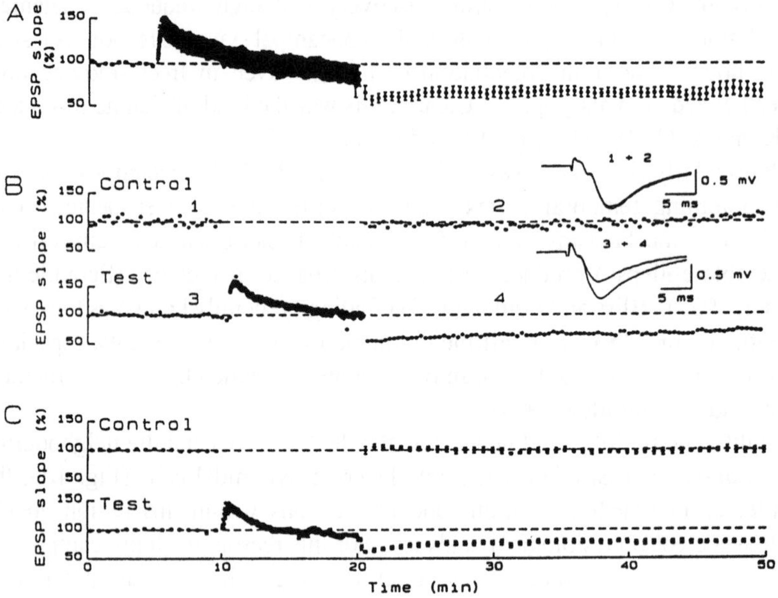

Figure 2. Prolonged low-frequency stimulation induces synapse-specific LTD. A: Summary of of data from 7 slices in which 1/s stimulation for 15 min resulted in 36% (SE ± 7%) reduction in initial slope of CA1 EPSPs. B: experiment in which two separate inputs onto same population of cells were stimulated alternatively (at 0.1/s), but only test pathway was stimulated at 1/s for 10 min. Inset are individual traces recorded at times indicated. C: Summary of 5 similar experiments. Note significant depression (by 21% ± 3%) only of test pathway (Fig. 1 from Mulkey and Malenka, 1992).

Over the last 20 years, a huge amount of effort and ingenuity has been devoted towards two principal questions. First, what cellular and molecular molecular mechanisms are involved in the induction and the maintenance of LTP? Second, does LTP really have anything to do with learning and memory? Attacking these questions clearly required quite different strategies and experimental approaches. As might be expected in view of the complexity of biological "material", though much has been learned, we still have only partial answers to these questions.

2. SYNAPTIC TRANSMISSION IN THE CENTRAL NERVOUS SYSTEM

The main features of transmission at central synapses can be summarized as follows.

Signalling between cells by purely electrical means (for example where adjacent cells make contacts via low resistances "gap" junctions, Bennett 2000) is by no means of negligible importance; but far more common is transmission mediated by a chemical agent: at most synapses, the excitatory transmitter is the dicarboxylic aminoacid glutamate. Typically, numerous axon terminals -- only ≈1 µm in diameter -- form synapses with the much larger cell body and its dendritic branches, especially on dendritic "spines". The presynaptic terminal is specialized to manufacture and store the transmitter (eg. glutamate) in synaptic vesicles (≈50 nm) from which it is released when a vesicle fuses with the presynaptic membrane. It then rapidly diffuses into and across the minute synaptic cleft (much <1 µm) and binds to specific receptors in the post-synaptic membrane. In this way, glutamate causes a marked increase in Na^+ permeability; and by generating an influx of Na^+, it depolarizes the post-synaptic membrane -- this is the EPSP. If the EPSP is large enough, it triggers the voltage-dependent self-regenerative increase in Na^+ permeability that generates the action potential which travels along the axon and all its branches till it reaches the axon terminals at the synapses. By triggering an influx of Ca^{2+} into the terminal, the action potential sets into motion a complex sequence of events that lead to the fusion of the vesicle with the surface membrane and release of the transmitter into the synpatic cleft. The contents of one vesicle thus represent one quantum of transmitter (typically some 100 or more molecules). The potency of transmission is thus determined by the number of vesicles that release their quanta of transmitter (more or less simultaneously) -- this is a steep function (4th power) of the amount of Ca^{2+} entering the terminal. The second main determinant of potency is the affinity and density of post-synaptic receptors to which transmitter molecules bind.

A given cell is connected by thousands of synapses. Many of these synapses release an inhibitory transmitter -- by selectively increasing the permeability to Cl^- or K^+, inhibitory transmitters make depolarization more difficult and thus prevent excitation. Continually varying excitatory and inhibitory inputs, arriving over many different pathways, control the fluctuating state of excitability and firing of nerve cells throughout the brain.

3. MECHANISMS OF LTP

3.1 Induction of LTP

"Classical" LTP is most often induced by high-frequency ("tetanic") stimulation of a synaptic pathway. Because the crucial effect of such stimulation is strong depolarization of the post-synaptic cell, a similar LTP can also be induced by pairing *low-frequency* stimulation with direct depolarization of the post-synaptic cell by current injected through the recording electrode (as in the experiments illustrated in Figs. 7-9). Non-classical forms of LTP include "chemical" LTP: for example, LTP induced by depolarizing agents such the main excitatory transmitter glutamate, high [K^+] and K^+ channel blockers (Bliss and Collingridge, 1993); temporary replacement of medium glucose by

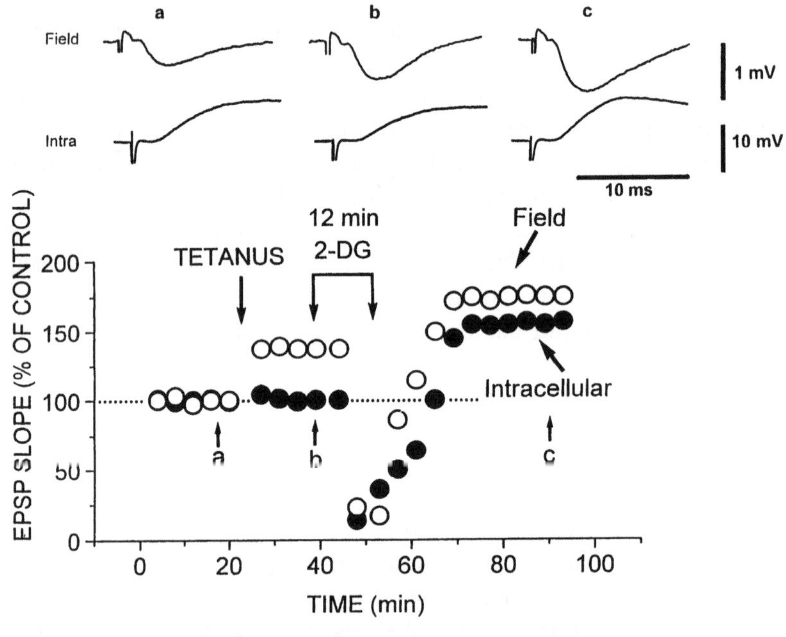

Figure 3. Intracellular EGTA prevents LTP induction by tetanic stimulation but not chemical LTP caused by temporary replacement of glucose with 2-deoxyglucose (2-DG). Simultaneous extra- and intracellular recordings were from CA1 in rat hippocampal slice; each point is average of 8 traces. Tetanic stimulation (100/s for 1 s, repeated once after 20 s) elicited usual LTP of extracellular (field) EPSPs (*open circles*) but no LTP of intracellular EPSP (*closed circles*). Note partial block of EPSPs after 12 min of 2-DG (10 mM, replacing glucose); then recovery, followed by similar potentiation of both field and intracellular EPSPs. Membrane potential was kept at -70 mV throughout. Inset traces were obtained at times indicated. Intracellular electrode contained 3 M KCl+100 mM EGTA (Fig. 2 from Zhao and Krnjevic, 2000).

equimolar 2-deoxyglucose produces an intriguing form of LTP (Tekkök and Krnjevic, 1995) not associated with post-synaptic depolarization (Zhao *et al.*, 1997) and not dependent on post-synaptic Ca^{2+} influx (Fig. 3).

With regard to classical LTP, there is a wide consensus on only two points.

A) *That classical LTP is initiated by* Ca^{2+} *influx into the post-synaptic cell.*

The crucial role of Ca^{2+} is clear. As first reported by Lynch *et al* (1983), tetanic stimulation fails to induce LTP in cells recorded with microelectrodes containing strong chelators such as EGTA (Fig. 3). Bearing in mind the essential role of Ca^{2+} as the internal messenger for many cellular processes (Berridge, 1997), that it triggers LTP might not seem surprising. Thus, in contrast to short-term synaptic facilitations -- including post-tetanic potentiation -- whose mechanisms are wholly presynaptic and likely caused by Ca^{2+} accumulation in the *presynaptic terminals* (Kamya and Zucker, 1994), LTP is induced post-synaptically, by an increase in cytoplasmic [Ca^{2+}].

B) *That the* Ca^{2+} *channels mainly responsible for the post-synaptic* Ca^{2+} *influx are of the NMDA receptor type.*

Several types of voltage-dependent Ca^{2+} channels exist, which are activated by membrane depolarization. The channels that mediate Ca^{2+} influx into nerve terminals are mainly of the N and P/Q-types. The main somatic channels that would be activated by depolarization are of the L-type.

Does the excitatory transmitter, glutamate, trigger Ca^{2+} by simply depolarizing the cell and thus opening such voltage-dependent Ca^{2+} channels? Things turned out to be more complicated. Glutamate binds to several different kinds of receptors. One, the AMPA-type (named after the more specific non-endogenous ligand AMPA) acts by enhancing Na^{+} influx: it is responsible for the fast excitatory post-synaptic potentials (EPSPs) observed at most synapses. Slower Na^{+}- and Ca^{2+}-mediated EPSPs are produced by glutamate acting on NMDA-type receptors: a crucial feature, most unusual for ligand-gated receptors, is that these receptors are strongly voltage-dependent; under resting conditions, they are blocked by Mg^{2+}, but this block is relieved by depolarization (Nowak *et al.*, 1984). If simple Na^{+}-mediated EPSPs cause sufficient depolarization to open NMDA receptors, the Ca^{2+} influx needed for LTP is generated.

As illustrated in Fig. 4, if NMDA receptors are blocked by a specific antagonist, LTP cannot be induced by tetanic stimulation -- but the same antagonist does not cancel an already established LTP. These remarkable receptors are thus an essential requirement for the induction of plastic changes at many synapses (one exception is the mossy fibre synapse between the axons of granule cells and the pyramidal cells of the CA3 region of the hippocampus (Bliss and Collingridge, 1993, Nicoll and Malenka, 1995)).

It should be noted that LTD also depends on Ca^{2+} influx via NMDA receptors (Linden and Connor, 1995). This may seem paradoxical. But it may

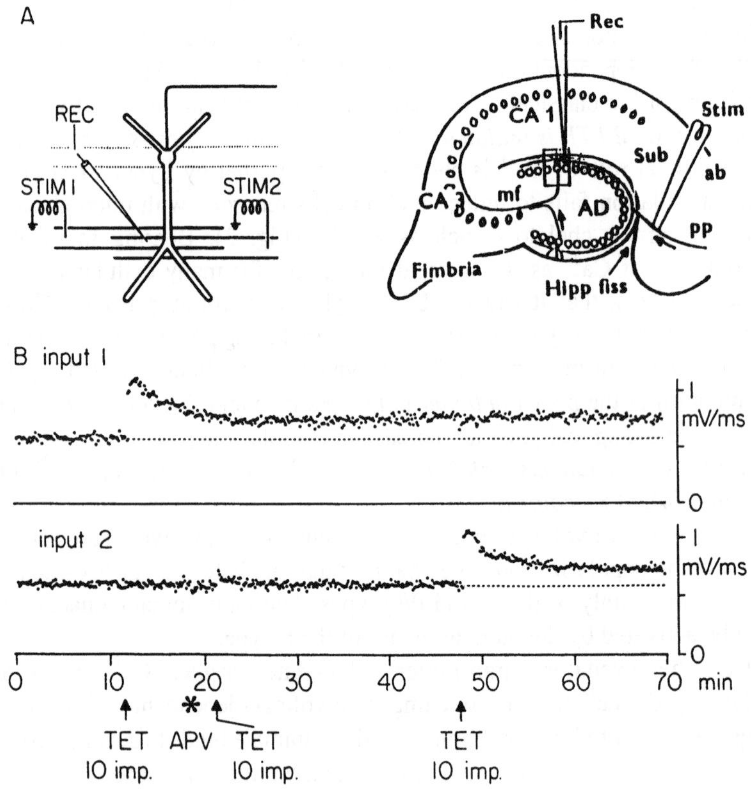

Figure 4. In CA1 region of hippocampal slice, LTP is pathway-specific and NMDA receptor-dependent. A: at right, diagram of hippocampus shows dentate area (AD) and CA3 and CA1, as well as location of stimulating electrode in perforant path (pp) and recording electrode in AD for experiment illustrated in Fig. 1; at left, diagram of pyramidal cell in CA1 with recording electrode in dendritic region where synapses are formed by afferent fibres -- two electrodes (stim 1 and stim 2) stimulate separate sets of fibres that make synapses on the same population of pyramidal cells. B: EPSPs generated by alternating stimulation of input 1 and 2; data plotted are initial slope of EPSPs which is a good index of efficacy of synaptic transmission. Throughout, EPSPs were evoked at 0.2 Hz. After initial control run, input 1 was tetanized with 10 impulses at 50/s (first arrow): this resulted in LTP of the tetanized synapses, but not of the other pathway. At star, NMDA antagonist aminophosphonovaleric acid (APV) was applied to the slice: this did not abolish the LTP of the potentiated input 1 but prevented any potentiation of input 2 when it was subsequently tetanized (second arrow); but after APV was washed-out, a second similar tetanus (third arrow) induced LTP of input 2 but had no effect on input 1 (Fig 1 from Wigström *et al.*, 1986; right half of A is Fig. 1A from Bliss and Lømo, 1973).

be explained if the direction of the plastic change is determined by the amount of cytoplasmic [Ca^{2+}] increase; probably even more important is the *pattern* of

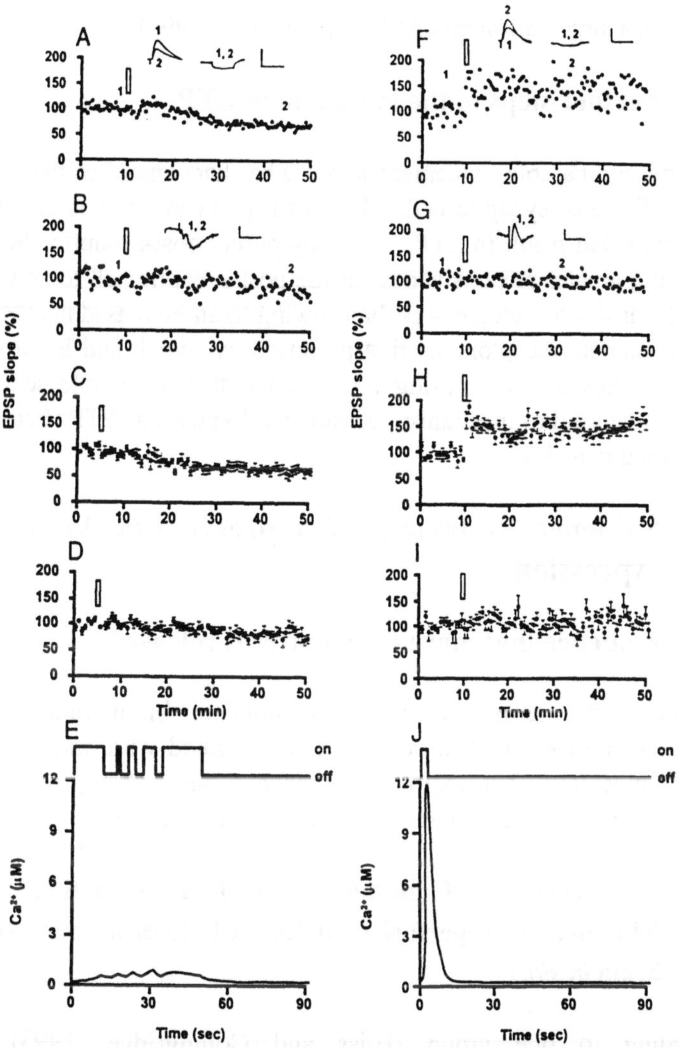

Figure 5. Direction of plastic change of hippocampal neurons depends on pattern of cytosolic $[Ca^{2+}]$ rise: modest and prolonged increases induce LTD, large sharp ones, LTP. A and F: whole-cell recording of EPSPs from neurons filled with light-sensitive Ca^{2+} buffer nitrophenyl-EGTA; inset are individual EPSPs and resistance-measuring pulses recorded at times indicated before release of caged Ca^{2+} (at vertical bar); time course of changes in EPSP initial slope is plotted below. Note LTD in A and LTP in F. B and G: extracellular field recordings of population EPSPs show no comparable changes. C and H are averaged intracellular data from 5 and 10 slices respectively, and D and I correponding averaged extracellular recordings. Calibration bars indicate 10 mV and 50 ms for intracellular and 0.2 and 10 ms for extracellular data. E and J: calculated (approximate) increase in $[Ca^{2+}]_i$ produced by photolysis using different patterns of light exposure (Fig. 2 from Yang *et al.*, 1999).

[Ca^{2+}] rise -- as shown in Fig. 5, a prolonged small increase of [Ca^{2+}] tends to induce LTD, a large and sharp one LTP (Yang *et al* 1999).

3.1.1 Further steps in the induction of LTP

There is no comparable agreement about what happens after the initial Ca^{2+} influx into the post-synaptic cell. The main point at issue is the site of the further events that result in LTP. Are they purely post-synaptic, the principal change being greater responsiveness of the post-synaptic cell to an unchanged amount of transmitter release -- perhaps owing to an increased number of post-synaptic glutamate receptors, as first proposed by Lynch and Baudry (1984)? Or are they presynaptic, causing more transmitter to be released from the terminals -- the kind of mechanism already well-known as PTP (Eccles, 1953), but much longer lasting?

3.2 Evidence in favour of a presynaptic locus of LTP expression

3.2.1 Evidence that more transmitter is released

Because nerve terminals in the mammalian brain (unlike some in invertebrates) are too small to be readily investigated with microelectrodes, a variety of ingenious, but indirect, approaches have attempted to measure synaptic glutamate release -- mostly with inconclusive results.

3.2.1.1 Measurements of glutamate in fluid contacting the brain (obtained by superfusion of isolated slices or microdialysis in brain *in situ*)

According to one group (Bliss and Collingridge, 1993), LTP is accompanied by an increase in ongoing and evoked release of glutamate. But this could not be confirmed by other authors (Aniksztejn *et al.,* 1989). In any case, it is unlikely that glutamate "overflow" can accurately reflect the small and transient release of glutamate that generates single EPSPs at the *potentiated* synapses (which make up only a fraction of all glutamatergic synapses in the tissue).

3.2.1.2 Monitoring of glutamate uptake by neighbouring astrocytes

A more promising approach is to record not from neurons but from adjacent non-neural satellite cells. After its release into the synaptic gap, glutamate is very efficiently removed from the extracellular space by active transport, mainly into the star-shaped astrocytes. Being driven by the Na^+ gradient and accompanied by net Na^+ influx, glutamate uptake is electrogenic; when recording from astrocytes, it is seen as an inward current that persists after glutamate receptor-mediated currents are pharmacologically suppressed (Diamond *et al.*, 1998; Luscher *et al.*, 1998). Because many synapses are fully ensheathed by astrocytes, changes in the magnitude of this current should reflect changes in glutamate release in the synapses. Conditions that potentiate transmitter release, such as low $[Mg^{2+}]_o$ and PTP, are indeed associated with larger glutamate transport currents; but no comparable change was seen during LTP (Diamond *et al.*, 1998; Luscher *et al.*, 1998). This seems rather convincing evidence that LTP is *not* caused by increased transmitter release. A possible objection, however, is that extensive electrical coupling between astrocytes could mask very localized changes in currents caused by increased glutamate release at a subset of synapses.

3.2.1.3 Use-dependent block of glutamate receptors monitors glutamate release

Agents that block only open channels progressively reduce EPSPs. An increase in glutamate release elicits more frequent activation of post-synaptic glutamate receptors. Therefore a use-dependent blockers should produce a more rapid depression of EPSPs if more glutamate is released after the induction of LTP. This ingenious technique, like most others in this field, has yielded conflicting results: no change (Manabe and Nicoll, 1994, Mainen *et al.*, 1998) or more rapid decrement of EPSPs (Kullmann *et al.*, 1996).

3.2.1.4 Quantal analysis of EPSPs

The quantal nature of transmitter release was first observed at the neuromuscular junction (Katz, 1971). In the absence of stimulation, spontaneous "miniature" synaptic potentials reflect the spontaneous fusion and emptying of single synaptic vesicles (single quanta). The size of an EPSP evoked by stimulation depends on the number of vesicles or *quanta* released. Any increase in release (or *quantal content*) can therefore be measured by comparing the amplitude of the evoked responses with that of the spontaneous "miniature" events (the *quantal size*). In addition, quantal analysis can be used to detect *post-synaptic* changes: if the post-synaptic membrane becomes more sensitive to the transmitter, the mean amplitude of the miniature events (the *quantal size*) should increase but the ratio of EPSP/mEPSP (the quantal content) would remain the same.

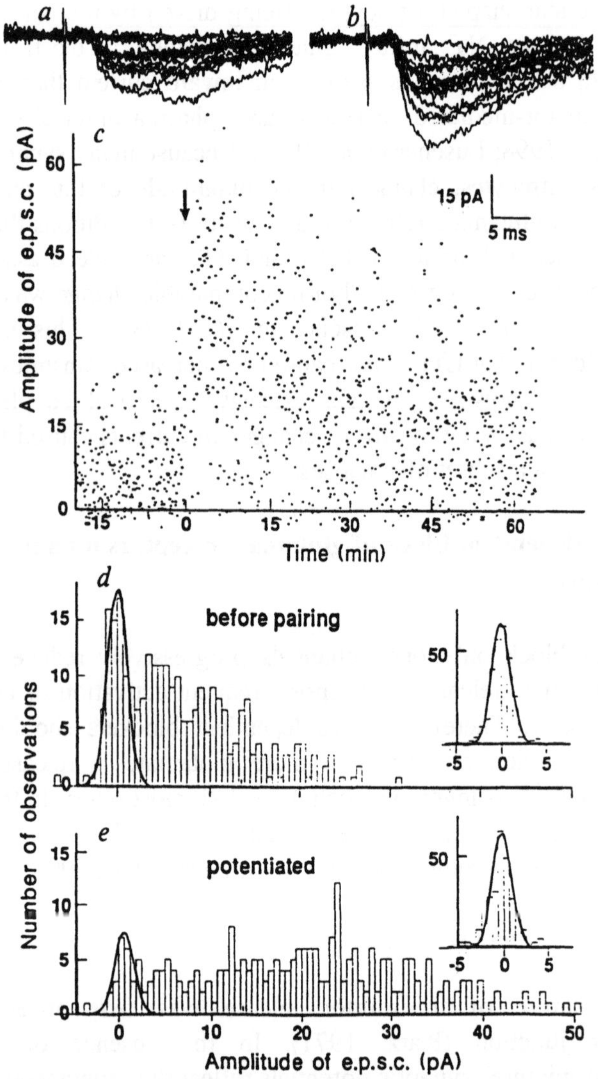

Figure 6. Reduced incidence of synaptic failures suggests a presynaptic site for LTP expression. Excitatory post-synaptic currents (EPSCs) were recorded from CA1 neuron under voltage-clamp (in hippocampal slice). a,b: superimposed traces of 16 consecutive EPSCs evoked by minimal stimulation, (a) before and (b) 40 min after LTP induction by pairing synaptic input with strong post-synaptic depolarization. c: dot-plot of minimal EPSC amplitude as function of time before and after pairing (at arrow): note over 2-fold increase in current amplitudes and their fluctuations, and 60% reduction in incidence of failures (traces with no detectable EPSC). d,e: histograms illustrate changed distribution of EPSC amplitudes after LTP; inset plots indicate no change in baseline noise (Fig. 4 from Malinow and Tsien, 1990).

Unfortunately, such quantal analysis cannot be easily applied to CNS synapses (Stevens, 1993). One difficulty is that any one cell receives numerous inputs from different neurons. All of them contribute to spontaneous minature EPSPs (mEPSPs) -- recorded in the presence of tetrodotoxin to eliminate spontaneous action potential-driven EPSPs -- but only a small proportion generate the EPSP evoked by stimulating a given pathway. Two indirect methods circumvent this problem. Both depend on the presumed Poisson or binomial character of the underlying random release process. For the first (the variance method), one measures fluctuations in amplitude of synaptic responses evoked by constant stimulation: if such fluctuations accurately reflect stochastic variations in quantal transmission, the mean quantal content is given by 1/(coefficient of variation)^2 ($1/CV^2$), which can be calculated from the mean size of the EPSPs and their variance. The second (the failures method) follows from another property of the Poisson distribution: if the probability of events is low, some of the stimuli fail to evoke any response at all, the incidence of such failures being given by exp(-m) (where m is the mean quantal content). Thus, the reduced incidence of failures after LTP seen in Fig. 6 is consistent with an increase in probability of release. Such evidence suggests that LTP is indeed expressed presynaptically by a lasting enhancement of evoked transmitter release (Bliss and Collingridge, 1993; Stevens and Wang, 1994; Kullmann and Siegelbaum, 1995).

Although quantal analysis proved of enormous value for studies of neuromuscular transmission, it has not fulfilled its promise in investigations of synapses in the brain: in the event, too many unsuspected factors complicated the interpretation of data to allow any definitive conclusions about the site of LTP expression.

3.2.2 Other evidence of a presynaptic change

One argument against a presynaptic locus of LTP is that EPSP "failures" may occur post- rather than presynaptically. It is based on the repeated finding that many synapses appear to have only NMDA receptors and therefore normally show no EPSP: after LTP, however, these synapses are no longer "silent", AMPA-type EPSPs becoming evident, presumably owing to the recruitment of AMPA receptors (Isaac *et al.*, 1995; Liao *et al.*, 1995).

3.2.2.1 Glutamate "spill-over"

Kullmann *et al.* (1996) proposed an alternative presynaptic explanation for "silent" synapses. Close examination of AMPA and NMDA receptor-mediated EPSPs and their variance, before and after LTP, showed clear increases in $1/CV^2$, strongly suggestive of increased quantal number, for both types of

EPSPs. To account for silent synapses, they proposed that glutamate may "spill-over" from active synapses to quiescent synapses on closely neighbouring cells: because NMDA receptors are much more sensitive than AMPA receptors, the glutamate spill-over would produce only NMDA-type responses, undetectable in normally polarized cells. With LTP, the enhanced release of glutamate would also activate the less sensitive AMPA receptors, and the synapses would no longer be silent at the normal potential of the cell.

3.2.2.2 Faster vesicle turnover

After they release their transmitter, synaptic vesicles reseal and are recycled for further transmitter uptake and release. While the interior of the vesicles is exposed at membrane surface, an intravesicular protein can be tagged with a fluorescent antibody. By monitoring vesicle turnover in this manner, Malgaroli *et al* 1995 observed a marked increase in vesicular recycling in LTP, which would be consistent with increased glutamate release. The main objection to this impressive achievement is that it was feasible only with cells in cultures.

3.2.2.3 Faster glutamate release

A different also presynaptic mechanism was recently proposed by Choi *et al* 2000. Using a rapidly reversible NMDA antagonist to assess the concentration of glutamate at silent synapses, they observed a large increase in cleft glutamate after LTP. In addition, applications of cyclothiazide -- an agent that enhances AMPA-type responses by suppressing their normally fast desensitization -- revealed slow AMPA-type EPSPs at "silent" synapses: these responses were much increased and accelerated after LTP. According to the authors, LTP might be caused not by an increase in glutamate release but rather by *faster* release through the fusion pore: hence the sharper rise and higher peak of glutamate concentration in the synaptic cleft.

3.2.3 Intercellular messenger

If LTP and LTD are expressed by a change in glutamate *release*, the induction process must somehow be transmitted from the post-synaptic cell, where it is initiated by a rise in [Ca^{2+}], to the relevant presynaptic terminals, across the synaptic gap. Essential requirements for the putative intercellular messenger therefore include: that it be rapidly produced by the rise in post-synaptic [Ca^{2+}], readily diffusible across cell membranes, but very localized in its action (which should not affect synapses on neighbouring cells). Arachidonic acid and nitric oxide (NO) are two of the proposed candidates that fit these requirements: both are produced by Ca^{2+}-sensitive enzymes (NO synthase and phospholipase A_2) and readily diffusible; and a possible role in LTP induction is indicated by some effects of blockers of their synthesis or action (Bliss and Collingridge, 1993).

3.3 Evidence in favour of a post-synaptic change

3.3.1 Smaller variance and LTP of NMDA component of EPSP

As new findings have been reported and fresh objections made to various aspects of quantal analysis, opinion has shifted back and forth between pre- and post-synaptic mechanisms. For example, the fact that NMDA receptor-mediated EPSPs show a lower variance and less LTP than do AMPA receptor-mediated EPSPs -- recorded in the same cell -- seemed to argue against a change in transmitter release, which ought to affect both more-or-less equally, and favour rather an increase in post-synaptic AMPA receptors (Kullmann and Siegelbaum, 1995).

3.3.2 mEPSPs recorded in presence of Sr^{2+} are larger after LTP component of EPSP

The amplitude of miniature responses (*quantal size*) normally cannot be used to detect changes in post-synaptic sensitivity because most of the recorded mEPSPs (or mEPSCs) originate from non-potentiated synapses. An ingenious way around this problem has been to record from slices bathed in medium containing Sr^{2+} (replacing the usual Ca^{2+}) (Oliet *et al.*, 1996): under these conditions, stimulation evokes EPSCs that are followed by bursts of mEPSCs arising only from the stimulated synapses (Fig. 7A). The fact that these miniature EPSCs became larger with LTP (Fig. 7B) and smaller with LTD (as usual, depending on the frequency of stimulation) strongly supported a post-synaptic site of change -- as would be expected if glutamate receptors were added to or were lost from the post-synaptic membrane.

3.3.3 Redistribution of post-synaptic AMPA recptors

Recent developments in optical techniques have made it possible to monitor with high resolution changes in post-synaptic glutamate receptors, especially those of the AMPA type. Using two-photon laser scanning microscopy, Shi *et al.* (1999) have been able to visualize fluorescently-labelled AMPA receptors even in slices. Under normal conditions, very few of the tagged receptors were at the tip of dendritic spines (where most of the synapses are situated), the great majority being within the dendrites. After tetanic stimulation – sufficient to

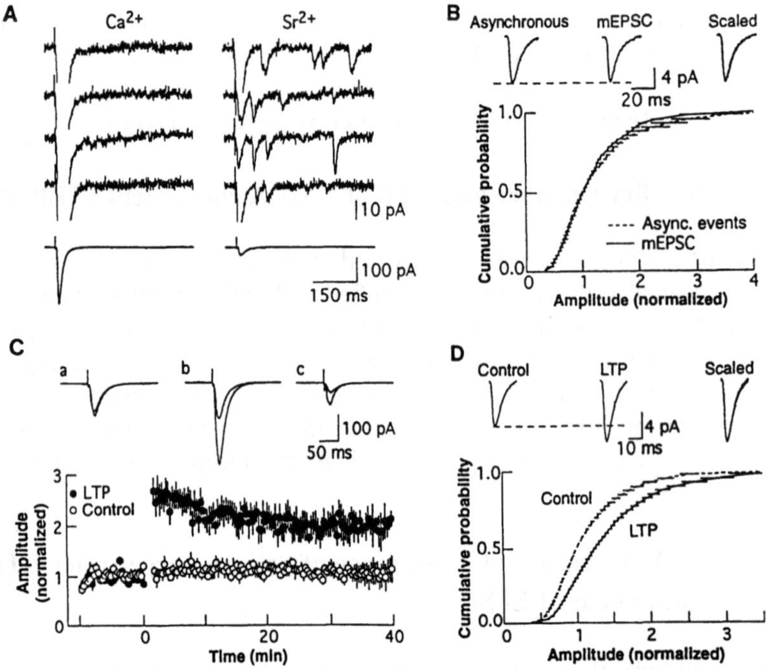

Figure 7. Larger unitary EPSCs after LTP indicate increased post-synaptic sensitivity to synaptic transmitter. When Ca^{2+} in external medium is replaced by Sr^{2+}, smaller EPSCs are evoked by stimulation, and they are followed by asynchronous discharge of miniature (unitary) EPSCs. A: samples of EPSCs recorded in CA1 neuron in presence of Ca^{2+} or Sr^{2+}; at bottom, averages of 30 traces at much lower gain; note much smaller response in presence of Sr^{2+}. B: plots showing similar cumulative amplitude distribution of spontaneous miniature EPSCs (mEPSCs) and asynchronous events recorded in Sr^{2+} (mean data from 6 neurons); above, averages of 100 such quantal events obtained in one cell do not differ in size or shape -- these data confirm identity between mEPSCs and asynchronous events. C: summary of six whole-cell recordings (in CA1) with stimulation of two separate synaptic pathways where strong depolarization of the post-synaptic neuron was paired with stimulation of one but not the other (control) synaptic input. Above are superimposed averages of 6 control and test responses obtained (a) before and (b) after LTP of test pathway, as well as (c) in presence of Sr^{2+}. D: Cumulative amplitude distribution of Sr^{2+}-induced asynchronous events obtained by stimulation of control and test pathway differ significantly ($P<0.0001$). Above are averages of 100 events recorded in one cell (Figs. 1A,B and 2A,B from Oliet *et al.*, 1996).

elicit LTP -- there was a marked redistribution of the tagged receptors, many now appearing in dendritic spines (including their surface), where they were previously either virtually absent (possibly "silent" synapses) or present in much lower number. Like the LTP, the redistribution of the tagged AMPA receptors was NMDA receptor-dependent. Opposite changes in AMPA receptors (that is, internalization of surface receptors) accompany LTD (Carroll *et al.*, 1999).

3.3.4 Role of protein kinases

What do we know about the events linking the post-synaptic rise in [Ca^{2+}] to the redistribution of AMPA receptors? Though other kinases (including protein kinases A and C and tyrosine kinase) may well play a significant part in the induction of LTP (Bliss and Collingridge, 1993; Lu *et al.*, 1998; Lee *et al.*, 2000), there is a broad consensus in favour of α-calcium-calmodulin-dependent protein kinase II (CAMKII) as a crucial link between Ca^{2+} and further events (Bliss and Collingridge, 1993; Lisman, 1994; Malenka and Nicoll, 1999; Soderling, 2000). This is indicated by several findings. CAMKII is present in large amount near the post-synaptic membrane. After LTP induction, there is a prolonged increase in its activity. Exogenous CAMKII inserted into cells enhances EPSPs (Fig. 8), whereas selective blockers of this kinase prevent LTP induction. An intriguing property of CAMKII that would promote long-term changes is self-regulation: once activity is initiated by Ca-calmodulin, the kinase is maintained in its active state by autophosphorylation. That this is important for LTP and other plastic changes has been demonstrated in a compelling way by introducing into the αCAMKII gene a point mutation at position 286, where threonine is normally autophosphorylated (Giese *et al.*, 1998; Glazewski *et al.*, 2000). In hippocampal slices from such mutants, LTP cannot be elicited. Moreover, these mice show impaired behavioural learning.

One post-synaptic component of LTP mediated by CAMKII is an increase in the unitary conductance of AMPA receptors (Benke, 1998; Derkach, 1999). Direct interactions between CAMKII and AMPA receptors have recently been described by Hayashi *et al.* (2000). Following up on their previous experiments with fluorescently-labelled receptors (Shi *et al.*, 1999), they have now shown that infecting cells with the catalytic domain of CAMKII increases both EPSPs and the insertion of AMPA receptors at the synapse. This process of insertion appears to require some additional proteins that have a PDZ domain, which are believed to control the assembly and localization of membrane proteins at synapses (Garner *et al.*, 2000). Cells infected with recombinant GLUR1 (AMPA subtype) receptor, with a point mutation at the putative PDZ binding site in the carboxyl-terminus, showed no lasting potentiation by co-expressed CAMKII or by tetanic stimulation (Hayashi *et al.*, 2000) (Fig. 9). These

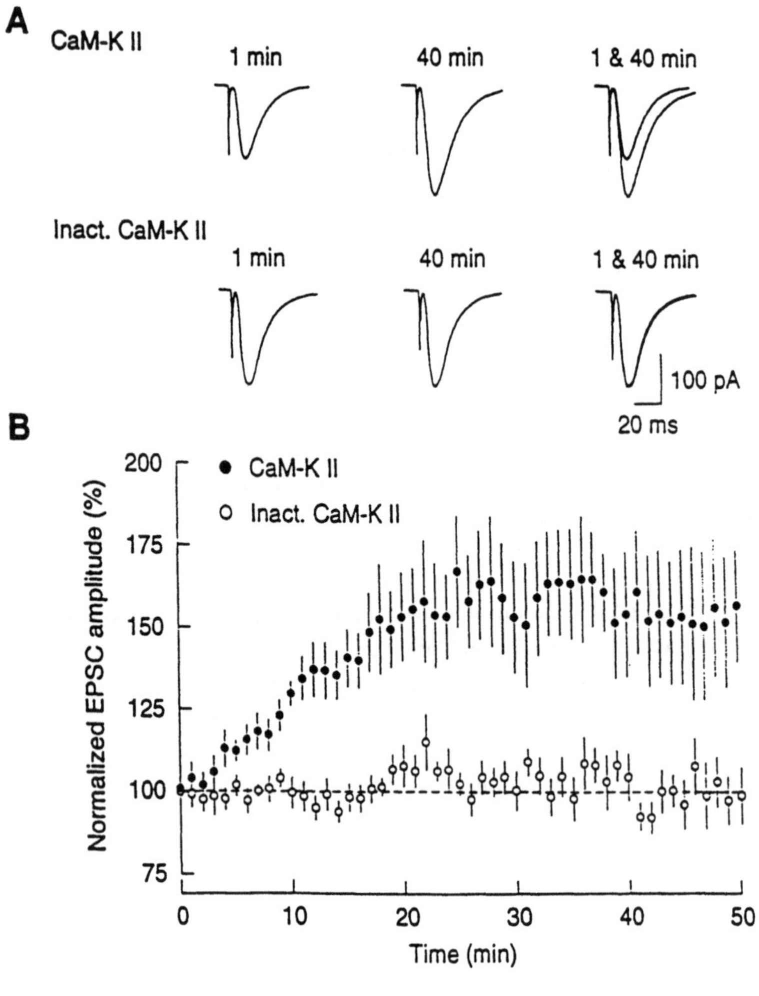

Figure 8. Activated form of CAMKII potentiates EPSCs of CA1 neurons. A: Average of six consecutive EPSPs at various times after start of whole-cell recording with electrodes containing active (upper traces) or heat-inactivated (lower traces) CAMKII. B: Mean time-course of changes in EPSC amplitude (± SEM; n=8 and 15 for upper and lower plots respectively) (Fig. 1 from Lledo *et al.*, 1995).

observations thus confirm previous evidence that, like LTP, CAMKII promotes the post-synaptic insertion of AMPA receptors; but this process requires an additional interaction between the receptors and the PDZ domain of some membrane proteins.

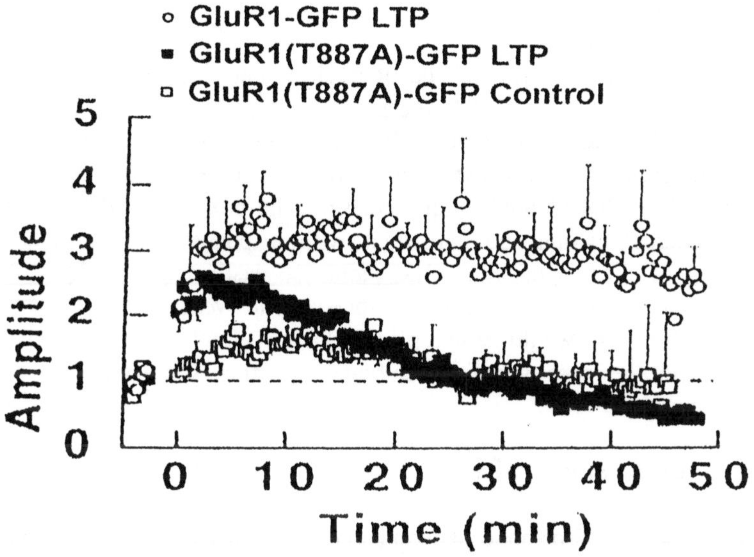

Figure 9. LTP requires delivery of AMPA-type glutamate receptors into post-synaptic membrane. To monitor insertion of AMPA receptors into membrane, GFP-tagged GLUR1 homomeric receptors were over-expressed in CA1 neurons in slices. In these cells (open circles), EPSCs showed sustained potentiation after coupling strong post-synaptic depolarization with synaptic stimulation (at time 0); such LTP was associated with the insertion of tagged GLUR1 into the membrane (not shown here). Because the subcellular localization of many surface proteins is controlled by an interaction between the cytosolic tail and proteins containing a PDZ domain, tagged GLUR1 mutated at this site (GLUR1(T887A)-GFP) was expressed in some slices. Such GLUR1 was inserted into the membrane and generated stable EPSCs (open squares, control pathway), but not sustained LTP (closed squares, test pathway). Thus, the participation of PDZ-domain proteins is essential for the insertion of AMPA receptors into the post-synaptic membrane, which is evidently one mechanism of LTP (Fig. 4D from Hayashi *et al.*, 2000).

3.3.5 Further evidence that LTP and LTD are mediated by redistribution of AMPA-type glutamate receptors

LTP is blocked if agents (such as NEM or botulinum toxin) that interfere with membrane fusion are injected into the *post-synaptic* cell (Lledo *et al.*, 1998). An interaction between one of the proteins to which such agents bind (NSF) and another subunit of AMPA receptors (GLUR2) appears to be important for LTD -- agents that specifically prevent or enhance this interaction can block or induce LTD (Lüthi *et al.*, 1999). These findings (and others (eg. Lee *et al.*, 2000)) suggest that LTP and LTD are not simply the opposite results of symmetrical phosphorylation and dephosphorylation of AMPA receptors, and that other sites and a variety of protein kinases are also involved.

3.4 Conclusion about pre- v. post-synaptic sites of LTP or LTD expression

In view of all the relevant data, it is clear that changes in numbers of functional post-synaptic AMPA receptors must play a major role in the expression of synaptic plasticity. But does this exclude any significant presynaptic changes? The answer is no. Admittedly, most of the evidence from quantal analysis that pointed to an increase in the probability of release (eg. fewer "failures") would be consistent with a post-synaptic mechanism, unmasking of silent synapses. One cannot so easily explain why LTP is accompanied by faster turnover of presynaptic vesicles. A reasonable conclusion therefore would be that plastic changes affect both the presynaptic terminal and the post-synaptic membrane -- either simultaneously or in succession.

3.5 Later stages of LTP

LTP is not an all-or-none phenomenon. It has at least three components (Bliss and Collingridge, 1993): a short-lasting STP (20-30 min), perhaps induced by relatively weak stimulation; an intermediate component that lasts at least some hours (and does not necessarily share the same mechanisms with STP (Schultz and Fitzgibbons, 1997)) -- this is the LTP seen in acute experiments, like most of those discussed in this review; and a much longer-lasting potentiaton, which probably depends on changes in cellular and synaptic structures that require the formation of new proteins -- as indicated by the absence of this late phase when LTP is induced after block of protein synthesis (Stanton and Sarvey, 1984) or in dendrites isolated from their cell bodies (Frey *et al.*, 1989). "Permanent" alterations in the number or structure of cells and synapses have been much discussed and investigated, but with rather few compelling positive results (eg. Geinisman *et al.*, 1996; Maletic-Savatic *et al.*, 1999). Nevertheless, one can assume that, albeit not conspicuous, a reorganization of existing synapses and/or growth of new structures, perhaps spread-out over a wide area, could mediate learning by providing a cellular basis for memory traces.

4. LTP AND LEARNING

Does LTP really represent more than an experimental artefact? One may well object that in many experiments, it has been induced by very unnatural tetanic stimulation of highly unphysiological preparations (slices or even cells in culture). On the other hand, LTP can be elicited by stimulating the tissue at frequencies close to those of physiological rhythms ("theta-burst stimulation" Bliss and Collingridge, 1993), as well as in the brain *in situ* (Bliss and Lømo, 1973) and even in freely moving animals (Bliss and Gardner-Medwin, 1973). Several properties of LTP make it rather suitable as a candidate learning mechanism: it requires the kind of association between post-synaptic firing and synaptic input postulated theoretically (Hebb, 1949). Albeit pathway-

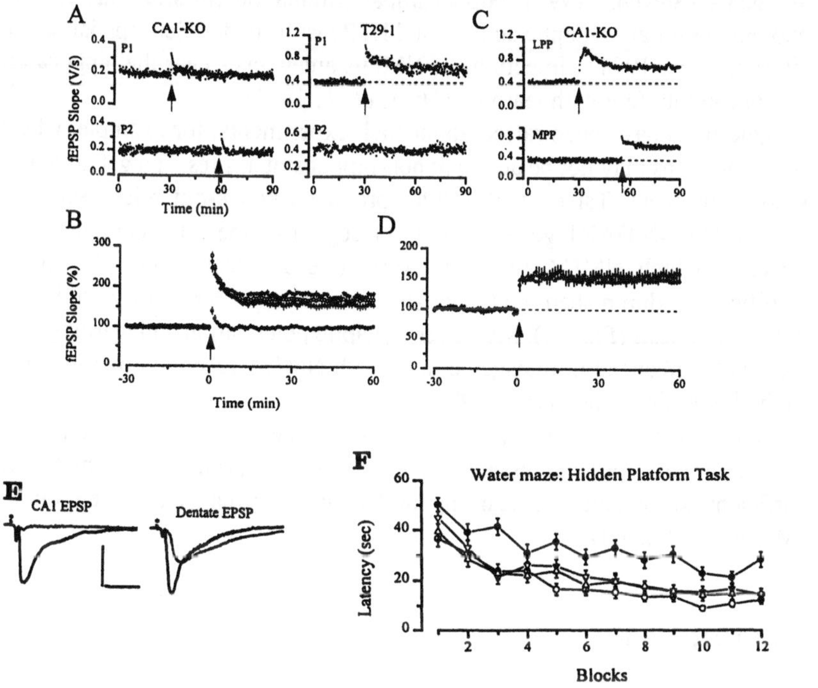

Figure 10. Spatial learning and hippocampal LTP are both defective in CA1-restricted NMDA receptor gene knock-out mice. A: in hipocampal slices from gene knockout (CA1-KO) and control (T29-1) mice, EPSPs were evoked in CA1 via two separate pathways (P1 and P2). Tetanic trains (100/s for 1 s, at arrows) induced LTP in the control but not in the CA1-KO slices; similar data from 21 CA1-KO and 21 control slices are summarized in B. C,D: LTP was unimpaired in similar tests on dentate gyrus, with independent stimulation of two different inputs (LPP and MPP, lateral and medial perforant pathways), showing that loss of LTP was indeed localized to CA1. E: superimposed EPSPs (in CA1 and dentate region of CA1-KO slice) were recorded first in standard medium (heavy trace) and then in presence of AMPA receptor antagonist CNQX (scales 0.5 mV and 10 ms); preservation of NMDA receptor-mediated component of EPSP in dentate but not in CA1 confirms selective deletion of NMDAR1 in CA1. F: Compared to control mice, CA1-KO mice are slower at spatial learning. Plots of the escape latencies of mice trained to find a hidden platform in a water maze by using distal cues. Note longer latencies for every block (4 trials per day) for CA1-KO mice (closed circles) than for three kinds of controls (open symbols) (Fig. 5A, Fig. 6 and Fig. 7A from Tsien *et al.*, 1996).

specific, it can affects very closely neighbouring inactive synapses (Engert and Bonhoeffer, 1997), which would permit associative, heterosynaptic learning); and though very long-lasting, it can be reversed by LTD.

Many studies have looked for more tangible evidence of a connection between LTP and behavioural learning. For example, clear mutual interactions (both positive by facilitation and negative by occlusion) have been reported between fear-evoked conditioning and LTP in pathways afferent to the amygdala (Stevens, 1998). Hippocampal stimulation (*in situ*) that is strong enough to saturate synaptic activity and LTP results in deficient spatial learning (Barnes *et al.,* 1994). Injections of NMDA antagonists near the hippocampus *in situ* can interfere with learning (Davis *et al.,* 1992).

Questions have been raised about such experiments: for example, whether such injections may not be acting at non-hippocampal sites. To get around this kind of problem, Tsien *et al.* (1996) produced knock-out mice selectively lacking the NMDAR1 gene in the CA1 region of the hippocampus. These mice, in which NMDA receptors were indeed defective in CA1, showed significantly diminished LTP (in slices) and a poorer ability to learn in behavioural tests (Fig. 10). Moreover, synaptic potentiation and learning ability were both enhanced in transgenic mice with NMDA receptors over-expressed in the forebrain (Tang *et al.,* 1999).

In conclusion, we now have good reason to believe that synaptic plasticity is a significant, perhaps essential component of learning; but whether it is sufficient to account for learning and memory remains an open question (Martin *et al.,* 2000).

REFERENCES

Aniksztejn, L., Roisin, M.P., Amsellem, R. and Ben-Ari, Y., 1989, Long-term potentiation in the hippocampus of the anaesthetized rat is not associated with a sustained enhanced release of endogenous excitatory amino acids. *Neuroscience* **28**: 387-392.

Barnes, C. A., Jung, M. W., McNaughton, B. L., Korol, D. L., Andreasson, K., and Worley, P. F., 1994, LTP saturation and spatial learning disruption: effects of task variables and saturation levels. *J. Neurosci.* **14**: 5793-5806.

Benke, T.A., Lüthi, A., Isaac, J.T. and Collingridge, G.L., 1998, Modulation of AMPA receptor unitary conductance by synaptic activity. *Nature* **393**: 793-797.

Bennett, M.V., 2000, Electrical synapses, a personal perspective (or history). *Brain Res Rev.* **32**:16-28.

Berridge, M.J., 1997, Elementary and global aspects of calcium signalling. *J. Physiol.* **499**: 291-306.

Bliss, T. V. and Collingridge, G. L., 1993, A synaptic model of memory: long-term potentiation in the hippocampus. *Nature* **361**:31-39.

Bliss, T.V. and Gardner-Medwin, A.R., 1973, Long-lasting potentiation of synaptic transmission in the dentate area of the unanaestetized rabbit following stimulation of the perforant path. *J. Physiol. Lond.* **232**: 357-374.

Bliss, T.V. and Lømo, T., 1973, Long-lasting potentiation of synaptic transmission in the dentate area of the anaesthetized rabbit following stimulation of the perforant path. *J. Physiol. Lond.* **232**: 331-356.

Carroll, R.C., Lissin, D.V., von Zastrow, M., Nicoll, R.A. and Malenka, R.C., 1999, Rapid redistribution of glutamate receptors contributes to long-term depression in hippocampal cultures. *Nature Neurosci.* **2**: 454-460.

Choi, S., Klingauf, J. and Tsien, R.W., 2000, Postfusional regulation of cleft glutamate concentration during LTP at 'silent synapses'. *Nature Neurosci.* **3**: 330-336.

Davis, S., Butcher, S.P. and Morris, R.G., 1992, NMDA receptor antagonist D-2-amino-5-phosphonopentanoate (D-AP5) impairs spatial learning and LTP in vivo at intracerebral concentrations comparable to those that block LTP in vitro. *J.Neurosci.* **12**: 21-34.

Derkach, V., Barria, A. and Soderling, T.R., 1999, Ca^{2+}/calmodulin-kinase II enhances channel conductance of alpha-amino-3-hydroxy-5-methyl-4-isoxazolepropionate type glutamate receptors. *Proc. Natl. Acad. Sci. USA* **96**: 3269-3274.

Diamond, J.S., Bergles, D.E. and Jahr, C.E., 1998, Glutamate release monitored with astrocyte transporter currents during LTP. *Neuron* **21**: 425-433.

Eccles, J.C., 1953, *The neurophysiological Basis of Mind.* Clarendon Press, Oxford.

Engert, F. and Bonhoeffer, T., 1997, Synapse specificity of long-term potentiation breaks down at short distances. *Nature* **388**: 279-284.

Frey, U., Krug, M., Brodemann, R., Reymann, K. and Matthies, H., 1989, Long-term potentiation induced in dendrites separated from rat's CA1 pyramidal somata does not establish a late phase. *Neurosci Lett* **97**: 135-139.

Garner, C.C., Nash, J. and Huganir, R.L., 2000, PDZ domains in synapse assembly and signalling. *Trends Cell Biol.* **10**: 274-280.

Geinisman, Y., Detoledo-Morrell, L., Morrell, F., Persina, I.S., Beatty, M.A., 1996, Synapse restructuring associated with the maintenance phase of hippocampal long-term potentiation. *J Comp Neurol.* **368**: 413-423.

Giese, K.P., Fedorov, N.B., Filipkowski, R.K. and Silva, A.J., 1998, Autophosphorylation at Thr286 of the alpha calcium-calmodulin kinase II in LTP and learning. *Science* **279**: 870-873.

Glazewski, S., Giese, K.P., Silva, A. and Fox K., 2000, The role of alpha-CaMKII autophosphorylation in neocortical experience-dependent plasticity. *Nature Neurosci.* **3**: 911-918.

Hansel, C., Artola, A. and Singer, W., 1997, Relation between dendritic Ca^{2+} levels and the polarity of synaptic long-term modifications in rat visual cortex neurons. *Eur. J. Neurosci.* **9**: 2309-2322.

Hayashi, Y., Shi, S.H., Esteban, J.A., Piccini, A., Poncer, J.C. and Malinow, R., 2000, Driving AMPA receptors into synapses by LTP and CaMKII: requirement for GluR1 and PDZ domain interaction. *Science* **287**: 2262-2267.

Hebb, D.O., 1949, *The Organization of Behavior*, John Wiley & Sons, New York.

Isaac, J.T., Nicoll, R.A. and Malenka, R.C., 1995, Evidence for silent synapses: implications for the expression of LTP. *Neuron* **15**: 427-434.

Kamiya, H. and Zucker, R.S., 1994, Residual Ca^{2+} and short-term synaptic plasticity. *Nature* **371**: 603-606.

Katz, B., 1971, Quantal mechanism of neural transmitter release. *Science* **173**: 123-126.

Konorski, J., 1948, *Conditioned Reflexes and Neuron Organization*, University Press, Cambridge.

Kullmann, D.M., Erdemli, G. and Asztely, F., 1996, LTP of AMPA and NMDA receptor-mediated signals: evidence for presynaptic expression and extrasynaptic glutamate spill-over. *Neuron* **17**: 461-474.

Kullmann, D.M. and Siegelbaum, S.A., 1995, The site of expression of NMDA receptor-dependent LTP: new fuel for an old fire. *Neuron* **15**: 997-1002.

Lee, H.K., Barbarosie, M., Kameyama, K., Bear, M.F. and Huganir, R.L., 2000, Regulation of distinct AMPA receptor phosphorylation sites during bidirectional synaptic plasticity. *Nature* **405**: 955-959.

Liao, D., Hessler, N.A. and Malinow, R., 1995, Activation of postsynaptically silent synapses during pairing-induced LTP in CA1 region of hippocampal slice. *Nature* **375**: 400-404.

Linden, D.J. and Connor, J.A., 1995, Long-term synaptic depression. *Annu. Rev. Neurosci.* **18**: 319-357.

Lisman, J., 1994, The CaM kinase II hypothesis for the storage of synaptic memory. *Trends Neurosci.* **17**: 406-412.

Lledo, P.M., Hjelmstad, G.O., Mukherji, S., Soderling, T.R., Malenka, R.C. and Nicoll, R.A., 1995, Calcium/calmodulin-dependent kinase II and long-term potentiation enhance synaptic transmission by the same mechanism. *Proc. Natl. Acad. Sci. USA* **92**: 11175-11179.

Lledo, P.M., Zhang, X., Sudhof, T.C., Malenka, R.C. and Nicoll, R.A., 1998, Postsynaptic membrane fusion and long-term potentiation. *Science* **279**: 399-403.

Lu, Y.M., Roder, J.C., Davidow, J. and Salter, M.W., 1998, Src activation in the induction of long-term potentiation in CA1 hippocampal neurons. *Science* **279**: 1363-1367.

Luscher, C., Malenka, R.C. and Nicoll, R.A., 1998, Monitoring glutamate release during LTP with glial transporter currents. *Neuron* **21**: 435-441.

Lüthi, A., Chittajallu, R., Duprat, F., Palmer, M.J., Benke, T.A., Kidd, F.L., Henley, J.M., Isaac, J.T. and Collingridge, G.L., 1999, Hippocampal LTD expression involves a pool of AMPARs regulated by the NSF-GluR2 interaction. *Neuron* 24: 389-399.

Lynch, G., and Baudry, M., 1984, The biochemistry of memory: a new and specific hypothesis. *Science* **224**: 1057-1063.

Lynch, G., Larson, J., Kelso, S., Barrionuevo, G. and Schottler, F., 1983, Intracellular injections of EGTA block induction of hippocampal long-term potentiation. *Nature* **305**: 719-721.

Mainen, Z.F., Jia, Z., Roder, J. and Malinow, R. 1998, Use-dependent AMPA receptor block in mice lacking GluR2 suggests postsynaptic site for LTP expression. *Nature Neurosci.* **1**: 579-586.

Malenka, R.C. and Nicoll, R.A., 1999, Long-term potentiation--a decade of progress? *Science* **285**: 1870-1874.

Maletic-Savatic, M., Malinow, R. and Svoboda, K., 1999, Rapid dendritic morphogenesis in CA1 hippocampal dendrites induced by synaptic activity. *Science* **283**: 1923-1927.

Malgaroli, A., Ting, A.E., Wendland, B., Bergamaschi, .A, Villa, A., Tsien, R.W. and Scheller, R.H., 1995, Presynaptic component of long-term potentiation visualized at individual hippocampal synapses. *Science* **268**: 1624-1628.

Malinow, R. and Tsien, R.W., 1990, Presynaptic enhancement shown by whole-cell recordings of long-term potentiation in hippocampal slices. *Nature* **346**: 177-180.

Manabe, T. and Nicoll, R.A., 1994, Long-term potentiation: evidence against an increase in transmitter release probability in the CA1 region of the hippocampus. *Science* **265**: 1888-1892.

Martin, S.J., Grimwood, P.D. and Morris, R.G., 2000, Synaptic plasticity and memory: an evaluation of the hypothesis. *Annu. Rev. Neurosci.* **23**: 649-711.

Mulkey, R.M. and Malenka, R.C., 1992, Mechanisms underlying induction of homosynaptic long-term depression in area CA1 of the hippocampus. *Neuron* **9**: 967-975.

Nicoll, R.A. and Malenka, R.C., 1995, Contrasting properties of two forms of long-term potentiation in the hippocampus. *Nature* **377**: 115-118.

Nowak, L., Bregestovski, P., Ascher, P., Herbet, A. and Prochiantz, A., 1984, Magnesium gates glutamate-activated channels in mouse central neurones. *Nature* 307: 462-465.

Oliet, S.H., Malenka, R.C. and Nicoll, R.A., 1996, Bidirectional control of quantal size by synaptic activity in the hippocampus. *Science* **271**: 1294-1297.

Pavlov, I.P., 1927, *Conditioned Reflexes*, p. 377, Dover Publications, 1960, New York.

Schulz, P.E. and Fitzgibbons, J.C., 1997, Differing mechanisms of expression for short- and long-term potentiation. *J. Neurophysiol.* **78**: 321-334.

Scoville, W. B. and Milner, B., 1957, Loss of recent memory after bilateral hippocampal lesions. *J. Neurol. Neurosurg. Psychiat.* **20**: 11-21.

Shi, S.H., Hayashi, Y., Petralia, R.S., Zaman, S.H., Wenthold, R.J., Svoboda, K. and Malinow, R., 1999, Rapid spine delivery and redistribution of AMPA receptors after synaptic NMDA receptor activation. *Science* **284**: 1811-1816.

Soderling, T.R. 2000, CaM-kinases: modulators of synaptic plasticity. *Curr. Opin. Neurobiol.* **10**: 375-380.

Stanton, P.K. and Sarvey, J.M., 1984, Blockade of long-term potentiation in rat hippocampal CA1 region by inhibitors of protein synthesis. *J. Neurosci.* **4**: 3080-3088.

Stevens, C.F. and Wang, Y., 1994, Changes in reliability of synaptic function as a mechanism for plasticity. *Nature* **371**: 704-707.

Stevens, C.F., 1993, Quantal release of neurotransmitter and long-term potentiation. *Cell* **72** Suppl: 55-63.

Stevens, C.F., 1998, A million dollar question: does LTP = memory?. *Neuron* **20**: 1-2.

Tang, Y.P., Shimizu, E., Dube, G.R., Rampon, C., Kerchner, G.A., Zhuo, M., Liu, G., Tsien, J.Z., 1999, Genetic enhancement of learning and memory in mice. *Nature* **401**:63-69.

Tekkök, S. and Krnjevic, K., 1995, Long-term potentiation in hippocampal slices induced by temporary suppression of glycolysis. *J. Neurophysiol.* **74**: 2763-2766.

Tsien, J.Z., Huerta, P.T., Tonegawa, S., 1996, The essential role of hippocampal CA1 NMDA receptor-dependent synaptic plasticity in spatial memory. *Cell* **87**: 1327-1338.

Wigstrom, H., Gustafsson, B. and Huang, Y.Y., 1986, Mode of action of excitatory amino acid receptor antagonists on hippocampal long-lasting potentiation. *Neuroscience* **17**: 1105-1115.

Yang, S.N., Tang Y.G. and Zucker, R.S., 1999, Selective induction of LTP and LTD by postsynaptic $[Ca^{2+}]_i$ elevation. *J. Neurophysiol.* 81: 781-787.

Zhao, Y.T. and Krnjevic, K., 2000, 2-Deoxyglucose-induced long-term potentiation in CA1 is not prevented by intraneuronal chelator. *J. Neurophysiol.* **83**: 177-180.

Zhao, Y.T., Tekkök, S. and Krnjevic, K., 1997, 2-Deoxy-D-glucose-induced changes in membrane potential, input resistance, and excitatory postsynaptic potentials of CA1 hippocampal neurons. *Can. J. Physiol. Pharmacol.* **75**: 368-374.

Subject Index

Adsorption sum model, 56, 57, 59
adsorption, 46-60
  cooperativity, 58
  equilibrium constant, 56
  kinetics, 53, 56
  transport-limited, 48
albumin, 46, 65, 70, 100, 104, 140, 141
aldehydes, 80
α-cyano-4-hydroxy cinnamic acid (ACCA), 39
α-lactalbumin, 45
Alzheimer's disease, 143, 145
amide bands, 76
amine oxidase, 120
amino acids, 18, 32, 34, 65, 148, 185, 199, 224
amyloid aggregates, 145
angiotensin I, 34, 40
aniline, 120
antibody, 145, 152, 196, 201, 202, 216
  arrays, 198
  profile, 200
  specificity screening, 197
apoB-100, 75, 79, 81, 83, 85
apolipoproteins, 75, 77, 81, 86
arachidonic acid, 216
atomic force microscopy, 198, 202
artificial neural network (ANN), 160
atherosclerosis, 76, 79, 86, 87
ATPase, 21, 22, 25
ascorbate, 109, 111, 112, 113, 114
astrocytes, 212, 213

autocorrelation function, 188
autophosphorylation, 219, 225
autoradiography technique, 52
average of the corresponding similarity scores, (AVS_g), 159
axon, 207

Bacteriophage, 144
BEND algorithm, 172
biradical disulfides, 96
BLAST program, 156, 158
buoyancy, 14, 15

C18-silica gradient surface, 46, 52-57, 59
Ca^{2+}-sensitive enzymes, 207
calcium, 140, 219, 224, 225
  channel brocker, 125
  ion antagonist, 125
calmodulin, 219, 225, 226
carbohydrates, 3, 4
channel blockers, 208
charged residue model, 33
cholesterol, 67, 75, 77, 81, 83, 86
cholesteryl esters, 75
chromosome, 147, 178
centromere, 148
circular dichroism, 84
collision induced dissociation (CID), 32, 37, 42
concanavalin A, 79
conductance, 5, 6, 219, 224, 225
consensus bendability scale14, 178
coronary artery disease, 46, 83
cortex, 215, 206, 225
Creutzfeldt-Jakob disease (CJD), 145
cumulative probabilistic score (C_g), 160
cutinase, 64
cyclothiazide, 216
cytochrome C oxidase, 79
cytokines, 76, 131

Darwinian behaviour, 144
decyl-β-D-maltopyranoside, 20

detergent, 15, 16, 20, 21, 68, 72, 199
diacylglycerols, 64, 66
dialkylglycero-alkylphosphonic acid-*p*-nitrophenyl esters, 68
dialkylglycerophosphonates, 70, 72
Diels-Alder reaction, 122
detergent, 15, 16, 20, 21
Detrendet Fluctuation Analysis (DFA), 190
dichloromethane, 52
dihydroxy-benzoic acid (2,5-DHB), 40
diffusion coefficient, 14, 15
di-tert-alkylnitroxides, 100
dithiocarbamate, 98, 99, 131, 133, 135
DNA, 27, 36, 65, 144, 145, 148-153, 168, 172, 175-178, 180, 182-186, 192, 195, 196, 198, 199, 200, 201, 202
bendability, 3, 7, 8, 9, 11, 171
bending, 167
bending anisotropy, 173
binding proteins, 167
conformations, 167
infidelity, 150
polymerase, 150, 151
polymorphism, 167
replication, 149, 150
sequence, 149, 169, 191, 200
structure, 167
supercoiling, 170
templates, 149
vibrational properties, 174
DNaseI, 171, 172, 178
domain sequences, 156, 158
drug delivery systems, 95
Dulbecco phosphate buffer, 52

Electrophoresis, 180, 199, 201
electrostatic charges, 4, 5
electron paramagnetic resonance (EPR), 89-104, 106-109, 111-116, 118-126
of nitric oxide, 98
longitudinally detected, 95
of thiols, 95
endothelial derived relaxing factor (EDRF), 131
endotoxin, 101, 105, 131, 141
EPR, see electron paramagnetic resonance

EPSCs, see excitatory post-synaptic currents
EPSPs, see excitatory post-synaptic potential
erythrocytes, 96
excitatory post-synaptic currents, 215
excitatory post-synaptic potentials, 215-223, 225-227, 219, 220, 223
exemplar-based methods, 164
esterases, 63, 64, 65, 72

Failures method, 215
Fenton-type reaction, 114
Field-Cycled Dynamic Nuclear Polarisation (FC-DNP), 95
Field-Cycled Proton-Electron Double-Resonance Imaging (FC-PEDRI), 95
finite element analysis, 170
flavins, 119
fluorescein isothiocyanate isomer, 52
fluorescence, 46, 48, 52, 54, 59, 64, 65, 66, 71, 72, 74, 84, 144-150, 182, 198
  anisotropy, 70
  correlation spectroscopy, 144-146
  decay time, 68
  emission spectra, 68
  intensity profile, 52
  spectroscopy, 63, 68, 74
  total internal reflection, 46
fluorine-containing nitrone spin trap, FDMPO, 110
fluorogenic substrates, 64, 66
fluorophore, 65, 69
F-NMR, 108, 138
foam cell formation, 76
fractal dimensions, 186, 187, 189, 191, 192

Ganglia, 125
gel electrophoresis, 13, 41, 51, 71, 72, 184
gel permeation chromatography, 13, 20
genetic information, 144, 146-148
genomics, 201
glucose, 208, 227
glucose oxidase, 120, 121
glutamate, 140, 207, 207, 209, 212, 213, 216, 217, 221, 225, 226, 227
glutaraldehyde, 53
glutathione, 97, 100, 102
gradient surfaces, 46, 48, 52, 60

glycerolipids, 64, 66, 68
glycoproteins, 10, 18, 19, 28

Halophilic bacteria, 10
HDL, see high density lipoprotein
helical circle diagrams, 177
hemocompatibility, 46
hidden Markov model, 156
hippocampus, 204, 205, 206, 209, 210, 225, 226, 227
high density DNA arrays, 195
high density lipoproteins, 45, 75, 76, 83, 84
high throughput screening, 145
hippocampus, 134
HIV-1, 145, 146
homogeneous isotropic elasticity parameter, 170
human T-cell receptor, 181
human protein p52, 191
Hurst exponent approach, 186, 191, 192
Hurst's rescaled range analysis, 189
hydroperoxides, 80
hydroxyl radical, 116
hydroxylaminc, 110, 112, 113, 114, 115
hydroxylamines, 100, 101, 102
hydroxy-5-methoxy-benzoic acid (DHBs), 40
hydroxy picolinic acid, 40
hydroxyl radical, 119
hydroxylamine, 120, 122, 125, 126

Imidazoline, 90, 94, 102, 103
imino nitroxyl radical, INR, 98
immunoassays, 195, 202
immunoelectron microscopy, 79
immunohistochemistry, 198
immunosensor array, 198
infrared spectroscopy, 75, 85
  of lipoproteins, 75
  attenuated total reflectance (ATR), 84
  Fourier transfer infrared spectroscopy, 85
  polarized attenuated total internal reflection Fourier-transform, 85
ion evaporation model, 33
iron (II) citrate, 134

ischemic injury model, 115
isotopes, 29, 108
  in chemical kinetics, 2
isotopic shift, 78, 82

Kaolin, 92
kinases, 157, 219, 221, 227
kinetic proof-reading, 148

Lasting depression of transmission (LTD), 206
light scattering, 10
LDL, see low density lipoprotein
long-term potentiation (LTP), 205
low density lipoprotein, 75-86
  oxidative modification of, 76
Leveque equation, 48
light scattering, 65
lipase, 66, 65-74
  activity, 65
  human hepatic, 66
  pancreatic, 65
  structure-function relationships, 63
lipid peroxidation, 125, 127, 129
lipolytic enzymes, 63, 64, 66, 69, 72, 73, 74
  physical properties of, 68
lipophilicity, 90, 92
lipopolysaccharide, 131, 141
lipoxy radicals, 119
lipoprotein (a), Lp(a), 46
liposomes, 94, 100, 101
L-monomethylarginine, 135
low density lipoprotein, 45, 48, 51-60
  adsorption, 46
Lowry method, 51
Lyapunov exponents, 187, 191
lysozyme, 45

Macrophages, 76
magnetic susceptibility, 140
membranes, 9, 15, 16, 20-22, 26, 75, 79, 84, 85, 90, 93, 94, 101, 121
  protein, 7, 15, 20
  receptors, 45

major intrinsic protein, 20
MALDI-MS, see mass spectrometry
malignant melanoma cells, 96, 102
mass, 197, 199
  determination, 28, 36
  molecular, 13-17, 19-22, 28, 30, 36, 47-50, 53, 56, 59, 71, 75
  mono isotopic, 35
  monomer, 16, 17
  weight average, 15
  resolution, 28, 34
mass spectrometry
  Electrospray/Ionization (ESI), 19, 27
  Matrix-Assisted Laser Desorption/Ionization (MALDI), 27, 28, 30, 37, 40, 42, 43, 190, 199
master sequence, 144
matrix isolation, 39
Matthieu equations, 32
membrane, 182, 196, 197, 200, 201, 205, 207, 209, 213, 216, 217, 219, 220, 221, 222, 226, 227
  depolarization, 209
  post-synaptic, 207
  surface, 216
  potential, 208
memory-based
  approach, 162
  categorization strategies, 163
  learning paradigm, 162
$(MGD)_2$ Fe(II) NO complex, 137
microarrays, 195, 197, 201, 202
microscopy, 20, 78, 85, 172, 217
microtubules, 147
mini-kink model, 172
minimal protein identifier, 199
mismatch repair, 149, 152, 153
mitochondrial ATP-synthetase, 23
molecular error catastrophe theory, 150
molecular diagnostics of virus, 145
molecular repair systems, 148
monocytes, 76
MRI spin trapping, 140, 141
murine neuroblastoma, 96, 102
muscle contraction, 7
myoglobin, 35-38, 41, 42, 104

Nearest neighbor method, 160
negative thixotropy, 5
neuromuscular junction, 213
neurons, 203, 205, 211, 212, 215, 218, 220, 221, 225, 226, 227
neurotensin, 34, 35, 40
nicardipine, 126
nifedipene, 125
nimodipine, 126, 129
nisoldipine, 126, 127
nitric oxide, 131, 140, 141, 216
nitric oxide synthase (NOS),, 131
nitrobenzene, 120
nitrocellulose, 196, 200
nitronylnitroxyl radicals (NNR), 98
nitroreductase, 120, 126
nitrosometoxyamines, 100
nitroxyl radicals (NR), 89
  pH-sensitive, 92
NMR, see nuclear magnetic resonance
NMR spectroscopy, 108, 110, 112, 115
NMR spin trapping (ST-NMR), 108, 110, 115, 116
nonstationarity, 187, 188
nuclear magnetic resonance, 6, 11, 62, 107-116, 129, 138, 139, 161
nucleic acid sequence-based amplification (NASBA), 145
nucleic acids, 144, 164, 165, 182, 183, 184, 196, 200
nucleosome, 169, 171, 172, 180
number of significant similarities (NSD_g), 159
nylon, 196

Octadecyldimethylchlorosilane (ODS), 52
octadecyldimethylsilyl (C18) groups, 46
oleic acid, 122, 123, 125
oligonucleotide, 169, 174, 197, 201
origin of life, 9, 143
osmotic pressure, 6
oxidative stress, 95, 97, 105, 115
oxyhemoglobin, 98
paramagnetic adducts, 109, 112, 114
partial specific volume, 14-16, 18, 19
  apparent, 15

(PATIR-FTIR) spectroscopy, see infrared spectroscopy
peak width definition, 28
peptide mass map, 42
pentraxins, 18, 19, 25, 26
"picket-fence" model, 84
phase plane embedding, 190
phase space, 190
phase space embedding, 187
phenylazotriphenylmethane, 111
phenylhydroxylamine, 120
phenyl-N-*tert*-butylnitrone (PBN), 135
phospholipid, 75, 85, 92, 94, 100, 103
phosphorus containing nitrone spin trap, DEPMPO, 108
plasma proteins, 18, 26
plasmids, 170
plasminogen, 46
P-NMR, 108
Poisson distribution, 215
polyacrylamide, 41, 72, 184, 196, 198
polyacrylamide gel electrophoresis (PAGE), 41
polyacrylic acid, 4
polyelectrolytc, 4-7, 3-11, 56, 92, 104
  charge-carrying, 10
  dimensions, 6
  thermodynamics of, 6
poly-L-lysine, 196
polymerases, 151, 153
polymers, 27, 104
  charge-carrying, 2, 3, 8, 9, 10, 11
polymethacrylic acid, 4
polymorphism analysis, 200
polypeptides, 2
polypropylene, 170
polyunsaturated fatty acyl chains, 80
polyvinylidene difluoride, 196
post-tetanic potentiation, 205, 209
potentiometric titration, 5, 6
prebiotic synthesis, 9, 105
presynaptic terminals, 209, 216
prion, 145
proteasomes, 148, 153

protein, 51, 62-73, 75-86, 92, 95, 105, 139, 140, 142, 144-149, 150, 151, 153, 156, 157, 159, 164, 166, 167, 172-174, 177, 180, 181, 185, 186, 195-202, 207, 210, 212, 213, 216, 218, 219, 221, 222, 227
  aggregation of, 13-16, 18-26
  adsorption, 46, 50
  adsorption rate, 48
  alpha-helices, 177
  arrays, 196
  biosynthesis, 150
  chips, 196
  conformational changes, 52, 69
  diffusion coefficient, 47
  sequence, 155, 178, 182, 186
  surface mass concentrations, 50
  interfacial property of, 45
proteome, 155, 199
proteomics, 28, 195, 201
proton relaxation enhancement, 137, 141
protonation, 34, 36
  of peptide, 40
pyramidal cells, 215, 209, 210
pyrene, 65, 69-71
pyrenebutyl-hexylphosphonic acid *p*-nitrophenyl esters, 72
pyridine, 52

Quadrupole field, 31, 32
quadrupole mass filter, 30-32, 42
quantal, 215, 217, 218, 222, 227
  analysis, 215
  content, 215
  size, 215
quasispecies, 144, 146
quencher, 64, 65

Reactive oxygen species (ROS), 119
receptors, 207, 216-222, 224, 225, 227
  AMPA, 215
  glutamate, 212, 213
  NMDA, 209
  voltage-dependent, 209
resonance energy transfer, 64, 72
robotic technology, 196

Scale symmetry, 189
scanning force microscopy (SFM), 46, 58
sedimentation, 14, 15, 19, 21-26
  coefficient, 14
  equilibrium, 14
  rate of, 14
  velocity, 14, 23
self similarity scores, 156, 158
semiquinones, 119
serum screening, 200
sequence dependent anisotropic bendability (SDAB), 171
sequence space, 144
Shannon's theory, 143
silanization reaction, 52
similarity based predictions, 155
simple elastic models, 170
simple scoring method, 160
small angle X-ray scattering, 10
sodium dodecylsulphate, 199
sodium nitroprusside, 100
spin labeling, 89, 103, 106
spin trapped adduct $(MGD)_2$ Fe(II)-NO, 140
spin trapping, 89, 99, 104, 105, 106, 115
spinal, 215
stearic acid, 122
streptomycin, 150
substance P, 34, 35, 40
succinic acid, 40
superoxide, 103, 114, 115, 116, 119, 120, 127, 128
surface plasmon resonance, 197
synaptic neuronal messenger, 131
synaptic plasticity, 213

Tetradotoxin, 215
theta-burst stimulation, 223
tissue, 98, 101, 102, 202, 212, 222
  arrays, 198
trans-lesion DNA synthesis (TLS), 151
transmitter, 206-209, 212, 213, 215-218, 225, 226
triacylglycerols, 71
trinitrophenylamine, 65
tumor HeLa cells, 96

Ubiquitination, 148
ultrafiltration, 65
universal protein array system (UPA), 197

Variance method, 215
viruses, 144
vitamin C, 120
viscosity, 5, 14
voltage-clamp technique, 205

Wiener-Khinchin theorem, 188

X-ray crystallography, 3, 4, 10, 62, 66, 68, 169, 185
xenobiotics, 119

Young's modulus, 170, 174, 175

Zeolites, 92

MIX
Papier aus verantwortungsvollen Quellen
Paper from responsible sources
FSC® C105338

If you have any concerns about our products,
you can contact us on
ProductSafety@springernature.com

In case Publisher is established outside the EU,
the EU authorized representative is:
Springer Nature Customer Service Center GmbH
Europaplatz 3, 69115 Heidelberg, Germany

Printed by Libri Plureos GmbH
in Hamburg, Germany